SOTHIS PRESS

Sothis Press aims to completely reimagine the public domain book with an emphasis on universal access to captivatingly designed print editions. We believe that technology will enable every single book ever published to be perpetually accessible in beautiful print form. Sothis is beauty in ubiquity.

www.sothis.press
1 Waterhouse Square, London, ECIN 2ST

The Una-Flow Steam-Engine

Johannes Stumpf

SOTHIS PRESS

ISBN 978-1-0365-1689-5

The Una-Flow Steam-Engine
is in the public domain.

The Una-Flow Steam-Engine.

By

Prof. J. Stumpf,

of the Technische Hochschule, Charlottenburg.

With 250 Illustrations.

London.
Constable & Company Ltd.
10 Orange Str. WC.

New York.
D. van Nostrand Co.
25 Park Place.
1912.

Printed by R. Oldenbourg, Munich

Table of Contents.

Preface.

All new doctrines are viewed with suspicion until some inquiring mind finds out that there is really nothing fundamentally novel in them. The principles on which the doctrines are based are proved to be old, but the fact remains that the Application of the principles is new. In the case of the uni-directional flow engine, or as it has been called for brevity, the "una-flow" engine, the above remarks apply very fully. After successful trials with my various designs of una-flow engine, friends and critics have been very ready and kind in pointing out that "something very like it has been done before". The great point is that nothing *exactly like it* has been done and none of the previous attempts have been attended with anything like the practical success which has attended my introduction of the una-flow engine about three years ago.

Of the earlier attempts, preference ought possibly to be given to Mr. J. L. Todd, who not only took out several patents in Britain for his improvements, but made actual tests with his engine. It is interesting to note the lines of development which Mr. Todd followed. He started off in his early British patent No. 7301 of 1885, with a very clear statement of the thermal conditions prevailing in an engine cylinder when the inlet was at the ends and the exhaust ports arranged in the centre of the cylinder and controlled by the piston. The features he emphasised were the "hot" inlet and the "cold" exhaust. From the very commencement of his investigation, Todd showed the tendency to move along wrong lines. He suggests that the excessive compression will be minimized by heat given up to the cylinder walls during compression which walls, he infers in order to perform this function, should not be jacketted. In fact J. L. Todd went so far as to heat the cold end instead of the hot end by placing the steam chest over the exhaust belt. These suggestions show the initial error Todd made, and it was probably this initial error which led Todd to depart from the pure una-flow engine and pass over to what he called the "dual" exhaust engine, viz, the mixed una-flow and counter-flow engine. The actual locomotive test made with Todd's engine was with the "dual" exhaust type and not with the *pure* una-flow type.

Another important point to bear in mind is that some ten years elapsed between Todd's first proposal and investigation of the una-flow engine and his introduction of the dual exhaust engine. It would appear therefore, that Todd had a hard struggle with the pure una-flow before he abandoned it for the "dual" exhaust. I am not aware that the dual exhaust engine was adopted to any large extent.

Many other inventors appear to have been fascinated with the uni-directional flow type, but I need only refer to Dr. Wilhelm Schmidt of Wilhelmshöhe. Dr. Schmidt does not appear to have gone so far as Todd. Like Mr. Todd Dr. Schmidt proposed to admit steam at one end of the cylinder and exhaust it through ports controlled by the piston. He made some proposals which can scarcely be taken as serious and practical, however interesting they might be otherwise. For instance, he proposes an automatic valve engine which depends upon varying throttling action and varying piston velocities for its operation. Then again, he does not always abide by the use of the cold annular exhaust belt which is connected by wide ports to the engine cylinder. This is an all important factor in the una-flow engine. Todd was ahead of Dr. Schmidt in this respect. In fact in view of the information gleaned from practical tests made with una-flow engines according to my design, it would appear that Todd, although he never obtained great success, was working on more correct lines than Dr. Schmidt.

I have to thank my critics for pointing out to me the work of these inventors and investigators, but would state that my investigations were entirely independent. Probably if I had advised myself fully of the work done by these gentlemen, I might also have been led astray. My investigations, however, have been entirely untrammelled.

I shall now briefly state the lines followed by me in my investigations. I set out to do in *one cylinder* what is usually done in *several cylinders* and I resolved to do this after the manner of a steam turbine, where the steam goes in hot at one end and has its energy extracted as is passes axially, always in the same direction, to the cold exhaust. Tackling the problem along these lines resulted in the forms of una-flow engine described in the pages of this book. It led to the following basic principles: — cut off early, use of a large ratio of expansion — keep the hot end hot — and the cold end cold. I have not departed from the first principle and in fact have only been more and more convinced of its soundness when followed to its logical limits. My various designs all have the main object in view of satisfying these basic conditions in as full a manner as possible.

What are the facts about the una-flow engine? Briefly these. — I do in one una-flow cylinder, what others do in two or three counter-flow cylinders (compound or triple). The results in steam consumption are the same, if not better. The cost for building and lubricating the una-flow engine is much less.

Expounding ideas is one thing — convincing one's fellow men to a sufficient degree to influence them to carry out the ideas into practice, is an entirely different and very much more difficult thing. My thanks are due to Mr. Smetana, manager of the Ersten Brünner Maschinenfabrik of Brünn for being the first whose convictions led to action. To quote this gentleman's own reply to my contentions "Your arguments are good — that I cannot deny — so I propose we put them to the test".

So the first una-flow engine was built by the Erste Brünner Maschinenfabrikgesellschaft in Brünn (Austria) in accordance with my design. It proved to be a full success and had a steam consumption equal to that of a good compound engine.

The first una-flow locomotive was built by the Kolomnaer Maschinenbau-A.-G., to the order of Mr. Noltein, the well known manager of the Moscow Kazan Railway, and proved eminently satisfactory. Since the time of its introduction, I have had no reason to complain of the slowness in advance. In fact at the end of July 1911, there were engines with a total output of over half a million horse power working or in actual construction.

This rapid development has entailed a vast amount of work in adapting the una-flow engine to all kinds of purposes. I am much indebted to my assistants in helping me to tackle this work and I would also specially mention Mr. Rösler of Mülhausen in Alsace, Mr. Arendt of Saarbrücken and Mr. Bonin of Charlottenburg, all of whom rendered me valuable assistance for which I thank them.

I must also express my thanks to all those gentlemen who allowed themselves to be convinced to the extent of action. In addition to Mr. Smetana already mentioned I have to thank Mr. Noltein of the Moscow Kazan Railway, Mr. Hnevkovsky of Brünn, Mr. Lamey of Mülhausen, Geheimrat Müller of Berlin and Mr. Schüler of Grevenbroich, for their assistance and support in the early stages of the development of the una-flow engine.

I have finally to thank Mr. P. S. H. Alexander of Messrs Mathys & Co., of 43 Chancery Lane, London for his services in the translation and preparation of this work in English.

Charlottenburg, Germany.

Chapter I.

The general thermal and constructional features of the una-flow steam engine.

As the name indicates, the energy of the steam is extracted in the case of the "una-flow" or "unidirectional" flow steam engine without causing the steam to return on its path, that is to say the steam passes always in one direction through the steam cylinder. As shown in fig. 1, the working steam enters from below into the hollow cover, heats the surfaces of the cover and then passes by the valve in the upper part of the cover into the cylinder; the steam then follows the piston whilst giving up its energy and after it has been expanded, it passes out, at the end of the piston stroke, through the exhaust ports arranged in the middle of the cylinder and controlled by the piston. In ordinary steam engines on the other hand, the steam has a counter flow action, that is to say it enters at the cylinder head, follows the piston during the working stroke, and then returns with the piston on its return stroke to exhaust through ports opening near the cylinder head. The counter flow or reversal of the exhaust steam causes considerable cooling of the clearance surfaces owing to their contact with the wet exhaust steam. This cooling action results in considerable initial condensation when the boiler steam is again admitted to the cylinder at the next working stroke. In the una-flow engine, all cooling of the clearance surfaces is almost entirely avoided and hence cylinder condensation is to a great extent eliminated as also is the necessity of employing several expansion stages. Una-flow engines therefore may be made with a single expansion stage, whilst the steam consumption will not exceed that of compound and triple expansion steam engines.

By eliminating all cooling of the clearance surfaces by the exhaust steam a somewhat similar effect is obtained as is got by superheating. In the ordinary engine, superheating is employed to overcome the above mentioned difficulties caused by the cooling of the clearance surfaces. If now, this cooling is avoided, it would appear that all necessity for superheating the steam is removed.

The use of a ring of exhaust ports or slots in the cylinder enables the area of the exhaust passage to be made three times as great as the port-area obtained by the use of slide or other valves. The result of this large exhaust area is that the end pressure in the cylinder is that of the condenser, especially when the use of long and narrow pipe connections between the condenser and the cylinder is avoided. In other words, if the condenser is arranged close up to the cylinder and the exhaust passage has a large cross section, it is possible to bring the cylinder pressure down to that of the condenser. In order to form a proper idea of the dimensions of the exhaust ports, one should imagine a piston valve of the same size as the working piston and a valve casing of the same size as the

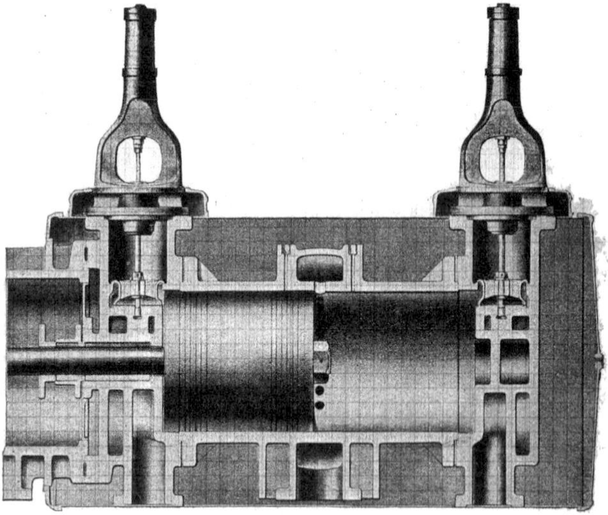

Fig. 1.

working cylinder, the piston valve being moved by an eccentric having the same throw as the engine crank. On an average, release takes place after $^9/_{10}$ ths of the stroke and consequently compression begins after $^1/_{10}$ th of the return stroke has been completed, or in other words, compression extends over $^9/_{10}$ ths of the stroke.

It will be evident that, by substituting exhaust ports or slots in the cylinder for the usual exhaust valve, all leakage losses at the exhaust valve and all the added clearance space and surfaces, which necessarily follow from the use of a special exhaust valve, are avoided.

The indicator diagram shows the expansion line to be an adiabatic for saturated steam and the compression line to be an adiabatic for superheated steam.

This is the best proof of the excellent thermal action of this engine. The excessive initial condensation, in an ordinary counter-flow engine using saturated steam, causes the expansion line to follow approximately the law of Mariotte. In the una-flow engine, using saturated steam, there is practically no initial condensation, so that the resulting expansion line is necessarily an adiabatic and all the more so if the steam is superheated.

Owing to the adiabatic expansion, the dryness fraction of the steam after expansion is very low. Thus in the case of steam having an initial temperature of 300 ⁰ C and an initial pressure of 12 atmos. expanding down to an end pressure of 0·8 atmos., the dryness fraction is 0·93, that is to say, the steam contains 7% of water. In reality the temperature of the steam at cut-off is liable to be a little less than the above owing to heat losses during admission. The result of these unavoidable heat losses during admission is that the expansion commences at a lower temperature and ends with a lower dryness fraction.

On the other hand, the heating jacket on the cover regenerates the steam during expansion. During expansion the cover jacket exercises considerable heating action owing to the considerable temperature difference between the cover and the steam, this heating action being transmitted principally to the steam immediately in contact with the cover. The steam immediately following the piston has a definite fall in temperature and increase in wetness owing to the adiabatic expansion. The greatest wetness is therefore found in the layer of steam immediately following the piston. In the layers between the piston and the cylinder cover, the wetness decreases until, in the case of the steam near the cover, some superheat may be present. Immediately on release, the wettest steam is forced out through the ring of exhaust ports in the cylinder walls. The steam which received heat during the entire time of expansion and exhaust and was subjected to the action of the full temperature difference between the expanding steam and the heating jacket is trapped by the piston and compressed, which compression will now approximate very closely to that of the adiabatic for superheated steam. This approximation to the adiabatic for superheated steam is still further assisted by the fact that during the first part of the compression, further heat is transmitted from the cover to the compression steam (fig. 2). Owing to the complete removal of all the moisture at each stroke, the well known heat losses, caused by the presence of water, in ordinary engines are avoided. Water hammer in the cylinder is in this way absolutely impossible.

Experimental investigation of steam jacketting on triple expansion engines shows, (I) in the case of a high pressure cylinder, no advantage, (II) in the case of the intermediate pressure cylinder, a small advantage, and (III) in the case of the low pressure cylinder, considerable advantage is obtained, in spite of the great losses which necessarily follow in the ordinary form of steam engine with a counter flow action of the steam. Counter-flow action necessarily involves the abduction of a considerable amount of heat from the jacket by the exhaust steam passing to the condenser. This will be appreciated when it is considered that, on the opening of the exhaust valve, a considerable amount of pressure energy in the steam is transformed into velocity energy, producing steam velocities in the ports and pipes between 350 and 400 metres per second. The wet exhaust steam sweeps over the clearance surfaces with this high velocity and deposits water of condensation on these surfaces. The result inevitably is that considerable re-evaporation takes place on account of the sudden fall in pressure and of the heat present in the walls of the clearances. The heat taken up by the clearances from the admission steam at each fresh charge is thus rapidly extracted during exhaust. A brief consideration will give a very fair idea of the uneconomical conditions in the ordinary steam engine, both as regards the loss of heat from the clearances and the loss of heat from jacketting. It should also be noted that the heat flow from the jacket

Fig. 2.

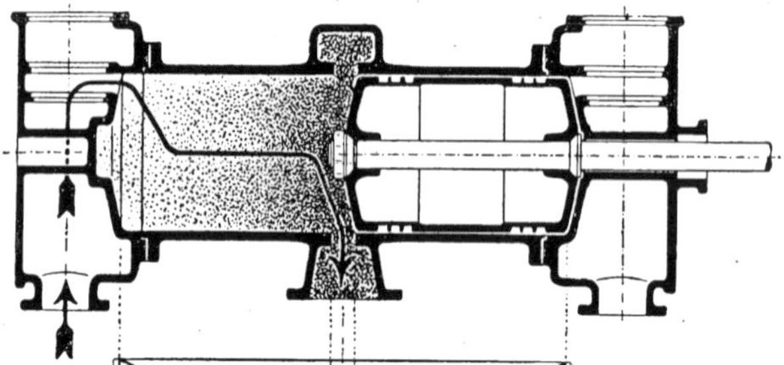

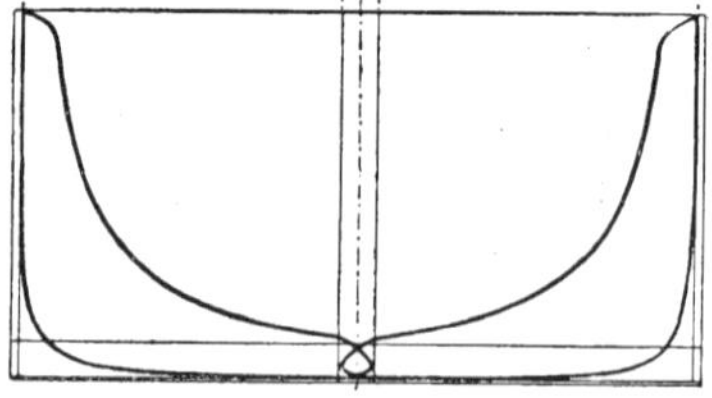

Fig. 3.

in an ordinary engine is greatest at the most unfavourable time, that is from the point where release commences to the point where compression begins or approximately during one half of a revolution, because it is during this time that the greatest temperature difference exists between the steam and the heating jacket. During the remaining half of the time of one revolution, the rate of heat flow from the jacket is less, as also is the velocity of flow of the medium over the hot surfaces. In spite of these disadvantages, the best results of jacketting are obtained in the low pressure cylinder. This may be explained by the fact that in the case of the low pressure cylinder, the heating jacket works with the greatest temperature difference. It follows from the above that in the case of a una-flow steam engine cylinder, a very efficient heating action must result as the heating in this case, as in the case of the low pressure cylinder of a triple expansion engine, works with the full temperature difference between the more or less fully expanded steam and the live steam. In addition to working with maximum temperature difference, it should be remembered that in the una-flow engine, the wasteful counter-flow of the ordinary engine is replaced by a uni-directional flow, so that not one single unit of heat is carried of from the jacketting by the exhaust steam passing through the exhaust ports. The exhaust steam, as a glance at fig. 2, will show, never passes over jacketted surfaces. The steam which comes in contact with the hot wall of the cover passes at most up to the neighbourhood of the exhaust ports, without however passing out through these ports. In consequence, the jacket heat can never be lost. The advantages of hot-jacketting the low pressure cylinder of a triple expansion engine must therefore be obtained in a much greater degree with a una-flow engine, because the great heat losses associated with the counter-flow action of the steam are wholly avoided by the new construction.

In the above, it has been assumed that the jacketting is limited to the cover or cylinder head, and that the cylinder proper is not at all jacketted (fig. 1 and 2). It is preferable to extend the cover jacketting to the point, where cut-off usually occurs, so that the clearance walls are most effectively heated on one side by the hot steam jacket and on the inside of the cylinder by the superheated compressed steam. The end temperatures which may be obtained will be more clearly realized from the following numerical example. — Dry saturated steam compressed from 0·05 atmos. absolute to 12 atmos. absolute, gives an end temperature of 807° C, according to the adiabatic for superheated steam. This example shows that for reducing the detrimental effect of the clearance surfaces it is not necessary to compress up to the admission pressure, but that a medium compression is quite sufficient. For reducing the detrimental effect of the clearance space it is best to make same so small that the compression runs up as high as the initial pressure, but for low condenser pressures it is mostly impossible to get the clearance space so small as to suit this requirement. (See chapter IX.)

It can be shown by comparison with van der Kerchove's construction that the una-flow engine is based on sound principles. The favourable results obtained by van der Kerchove are traceable to the special arrangement of the steam inlet valve in the cover and the heating action on the cylinder ends thereby secured. The advantages obtained by van der Kerchove are considerably enhanced in the case of the unaflow engine, owing to the working piston replacing the exhaust valve employed by van der Kerchove, and further, owing to the increased cover heating action and the reduced clearance spaces and surfaces resulting from the una-flow system. Kerchove's exhaust steam chamber at the end of his cylinder completely nullifies his hot steam chamber, and even goes further — it actually cools his admission steam, owing to its juxta-position to the steam chest.

The losses due to the clearance spaces and surfaces may be further reduced by reducing the metal thicknesses and by carefully machining the surfaces.

Although the entire cycle is carried out with steam in contact with the clearance surfaces, the cooling action in the una-flow engine is small, firstly because of the comparative stillness of the molecules of steam, owing to the exhaust outflow being relegated to the entirely opposite end of the cylinder space, secondly owing to the absence of any re-evaporation, and thirdly on account of the combined heating action of the cover jacketting and the high compression.

The thermal mix-up peculiar to the ordinary or counterflow engine is wholly avoided by the una-flow system. The cylinder consists really of two single acting cylinders set end to end with their exhaust ends common. The two diagrams are separated in proportion to the length of the piston. *The two ends are hot and remain hot, the common exhaust belt is cold and remains cold.* From the hot ends to the cold centre, there is, on either side, a gradual diminution of temperature which, when

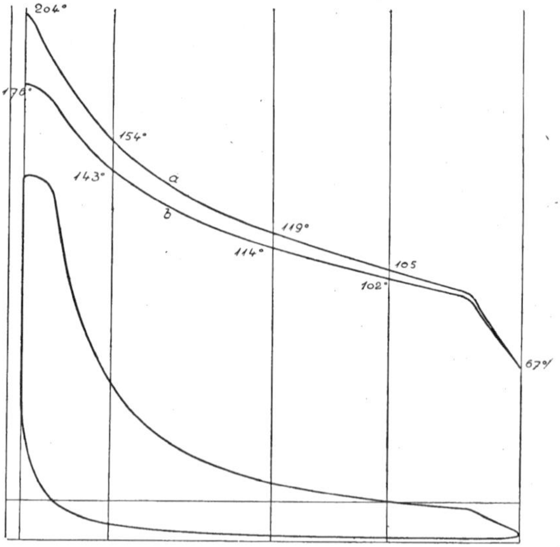

Fig. 4.

plotted out for each position of the piston, gives a line lying a little above the temperature line of the steam during admission and expansion. Fig. 4, shows the two lines plotted together, the upper one *a*, representing the temperature of the cylinder walls taken from actual measurements on a una-flow cylinder (non-jacketted cylinder, jacketted covers), whilst the lower one *b*, represents the steam temperature for the top or expansion line of the diagram.

This is in marked contrast to the ordinary or counter-flow engine where the two lines, as will be shown afterwards, have a very unfavourable situation and the exhaust end of one diagram projects pretty well into the inlet end of the other. In fact the entire engine is completely mixed up from a thermic point of view.

Owing to the disposal of the exhaust ports and the exhaust belt around the centre of the cylinder, the lowest temperature is secured in this part, where the piston velocity is greatest. This cooling is materially aided by dispensing with a jacket over the neighbouring parts of the cylinder. The piston has a large bearing surface and consequently a low specific bearing pressure. The cylinder is very simple in construction, being free from all cast-on casings or parts, so that local heating and casting difficulties may be practically eliminated, which results in the reduction or even elimination of internal strains and distortions. The large bearing surface of the piston, the cool exhaust belt and the simple form of the cylinder render the use of a tail rod quite unnecessary (see the designs of Gebr. Sulzer in Winterthur; Maschinenfabrik Grevenbroich; Globe-Iron Works in Bolton; Badenia in Weinheim; Gutehoffnungshütte). The piston is provided with two groups of rings, each group having usually three rings. Both groups of rings, that is usually six rings in all, are effective in packing the piston when the greatest pressure differences exist between inlet and exhaust. The steam pressure has usually fallen to about 3 atmos. when only a single group of rings is effective, that is when the other group has overrun the exhaust ports.

Experience with a large number of una-flow engines has shown that, even with very high superheating, the piston does not cause the slightest trouble if material and workmanship have been good and efficient lubrication is provided. Should the cylinder be scratched or scored owing to defective workmanship, bad material or bad lubrication, it is a very simple and quick operation to replace it, and owing to the simple form of the cylinder, the cost of such renewal is very low.

It is possible, without making any modification in the construction described, to work with superheats far in excess of those usually employed at present. Even with the highest initial temperatures the last part of the expansion line falls well below the saturation line, so that even with a hot steam jacket on the cover, reasonable and practicable working temperatures are obtained for the cylinder and piston.

Consequently the new construction opens up a further field of development by using higher temperatures. The new una-flow engine is all the more suitable for use with superheated steam as the superheat is favourable for the entire cycle, whilst in the ordinary counter-flow engine with multi-stage expansion, the superheat is excessive for the first cylinder and is too little in the succeeding cylinders. The excellent operation of the una-flow engine with superheated steam does not refute the fact, that it is also excellently suited for use with saturated steam. In reality the una-flow engine gives almost the same economic results with both kinds of steam. A further field of possibilities is opened up by the fact that higher initial boiler pressures may be employed. Pressures up to 30 atmos. should be quite as practicable in the una-flow engine with atmospherie exhaust as in the Diesel engine: A third field of development is opened up by combining high pressures and high temperatures.

The thermal, constructional and working advantages of the una-flow engine are such that the steam consumption for saturated and superheated steam is the same as that of good compound or triple expansion engines.

Chapter II.

The relation of the una-flow engine to the condenser.

A good condenser vacuum is of great advantage to the una-flow engine. The various compression lines for different initial pressures and the same end pressures are shown in figure 5, the clearance space being correspondingly altered for each case.

From these lines, it will be seen how the area of the diagram is considerably increased by employing a high vacuum. It is also possible, by suitable proportioning of the clearance space, to obtain any desired compression pressure and temperature. Working with a good vacuum, the clearance space may be very small, for instance the clearance space may be less than 1% of the working space of the cylinder with a vacuum of 0·05 atmos.

On comparing the duration of exhaust in the case of a una-flow engine, in which compression extends over $^9/_{10}$ths of the stroke, with the duration of exhaust in an ordinary counter-flow engine, it is found that the periods are approximately in the ratio of 1 to 2. The working steam in the una-flow engine must therefore be cleared out of the cylinder and passed to the condenser in one half of the time available in the case of the counter flow engine. It is a fact that in the present form of counter flow steam engine, a considerable pressure difference exists between the interior of the cylinder and the condenser, which pressure difference may be traced to the energy necessary to overcome the resistance of the ports and passages which are in almost all cases too narrow. As in the case of the una-flow engine the duration of the exhaust is reduced by one half, it is doubly essential to reduce, as far as possible, the resistance to the flow of steam to the condenser and this may be done by means of passages of large area and short length. These requirements are met to a considerable extent by the fact that the area of the piston-controlled exhaust ports in a una-flow engine may be three times as great as the exhaust port area of an ordinary counter-flow engine employing an exhaust valve. If the cross sectional area of the other passages are large in proportion to the large exhaust port area, and if the length of the connections from the exhaust port to the condenser is reduced as far as possible, all the requirements are satisfied for obtaining an end pressure within the cylinder equal to the pressure in the condenser. Experience proves this to be the case.

In figures 6 and 7, a una-flow engine is shown in cross section and longitudinal section. As can be seen the exhaust belt opens with its full breadth and diameter directly into the spray condensing chamber arranged below the cylinder. The spray is effected by means of a horizontal spraying tube introduced into the condenser chamber. As shown in the drawing, this construction provides an extremely large area for the flow of the exhaust steam whilst at the same time the length of the passages is reduced to the minimum. The result of this construction is equality in pressure within the cylinder and condenser when the exhaust ports are fully open.

In figures 8 and 9, a una-flow steam engine cylinder is illustrated in combination with a spray condenser of the Westinghouse-Leblanc type. The plant

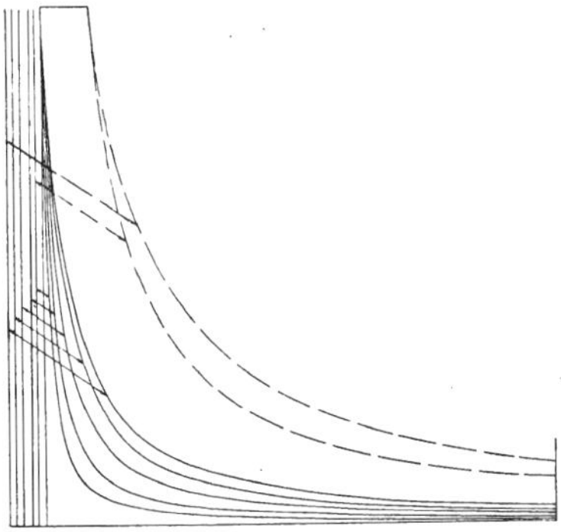

Fig. 5.

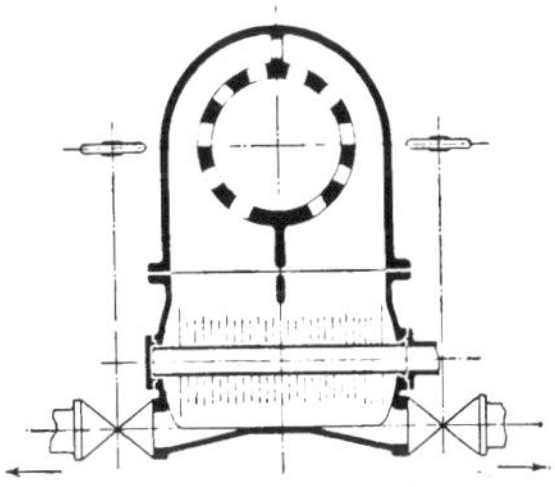

Fig. 6.

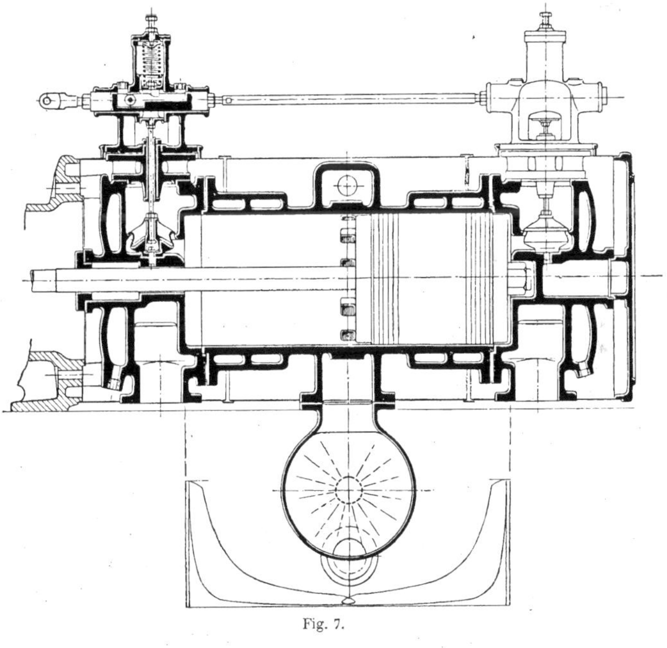

Fig. 7.

is so arranged that the condenser is also a supporting column for the una-flow engine cylinder. In this construction, as in the case of figures 6 and 7, a large cross sectional area is obtained for the flow of the exhaust steam, whilst the connections are extremely short. Here also there is complete equality between the end pressure of the steam in the cylinder and the condenser pressure. This complete equalisation of both pressures enables compression to start at the lowest possible point. The result is a considerable gain in the area of the diagram, a corresponding reduction of the clearance and the clearance surfaces, as well as a con-

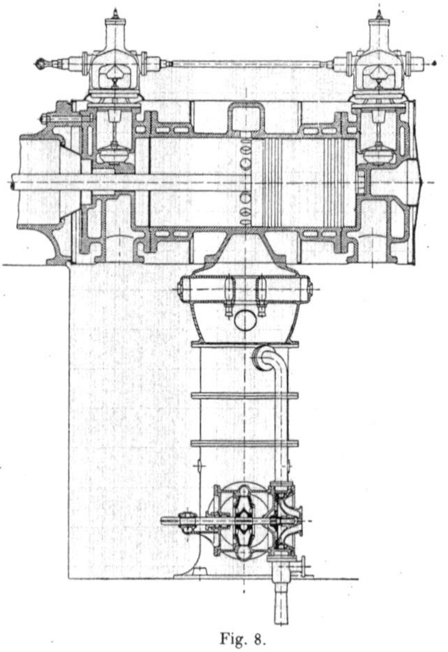

Fig. 8.

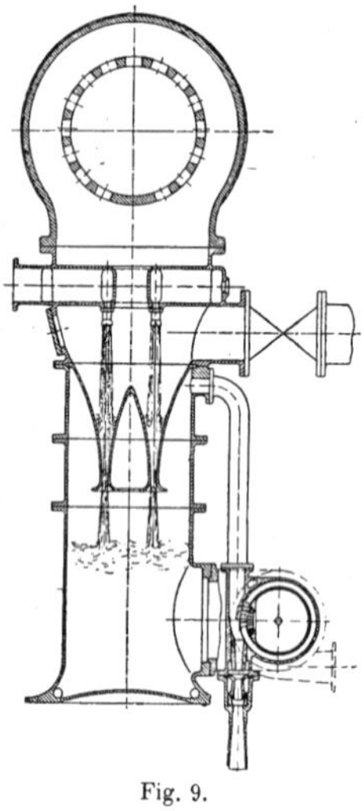

Fig. 9.

siderable improvement of the thermal efficiency (fig. 5). The reduction of the duration of the exhaust, with piston-controlled ports, also causes a corresponding reduction of the cooling of the cylinder when in communication with the condenser, as compared with ordinary counterflow engines. As soon as the exhaust ports are closed, the condenser cooling action ceases completely and during the entire remainder of the stroke the heating action of the jackets is effective. This heating action is not reduced or rendered inefficient by the nature of the exhaust or by any circumstances connected with the exhaust.

It is very bad practice to introduce anything in the nature of an oil separator, switching valve, reheater or bends in the connection between a una-flow steam engine and the condenser. Such parts cause very considerable resistance and should therefore be wholly avoided, if their introduction is not rendered absolutely necessary by other very important circumstances. The ports may also be formed as nozzles as shown in figure 10, in which case the nozzles are in the form of pipes opening into a steam turbine. This steam turbine would then work with intermittent impulses or puffs at each exhaust from the cylinder, the pressure-energy remaining in the steam being transformed into velocity-energy and extracted in the turbine. In this way a considerable part of the energy, represented by the toe of the diagram, which is cut off, may be usefully employed. A very good construction would be to employ three rows of exhaust ports, the two outer rows

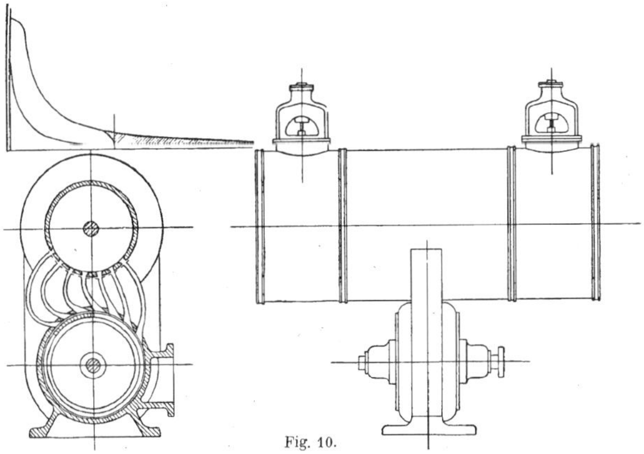

Fig. 10.

being connected as described to a turbine and the centre group of ports being connected directly to the condenser. The direct connection of the middle row of ports to the condenser produces the highest possible vacuum in the cylinder.

The connections to the atmosphere should be by way of the condenser. In such a case when the air pump, which is also connected up to the condenser, is shut down, the condenser will act as a silencer (see fig. 6).

Chapter III.

The steam jacketting.

During expansion, a very active transference of heat takes place between the cover and the layer of steam immediately in contact therewith, in consequence of the great temperature difference between the jacket and the expanding or expanded steam. The steam entering the cylinder contains less heat than the jacket steam and becomes wetter by the expansion. The greatest amount of condensation takes place in the layer of steam immediately following the piston. The moisture of the steam decreases in the layers between the piston and the cover, whilst that layer immediately in contact with the cover may at the end of expansion be dry or even superheated. During the exhaust, the wet steam is passed out of the cylinder through the ports. The quantity of steam therefore which, during the entire period of expansion, has been receiving heat from the cover at the full temperature difference between the expanded steam and the heating jacket, is trapped by the piston. This trapped residual steam is used for compression, and the compression will approximate very closely to the adiabatic for superheated steam. This approximation is still further promoted by the fact that during the first part of the compression more heat is given up by the cover to the steam being compressed. Experiments have shown that jacketting increases in importance as the temperature of the working steam approaches the saturation temperature. The importance of the jacketting therefore decreases proportionately as the steam is superheated. This will be seen from figure 11, which shows a unaflow steam engine cylinder with suitable jacketting for working with steam of a temperature of 350° C and over. In this case the jacketting is confined wholly to the cover. In figure 12, the construction of the jacket is illustrated for an engine working with steam at a temperature of 250° to 350° C, and in figure 13, the corresponding construction for saturated steam is given. On comparing the three figures, it will be seen that the jacket is extended from the end towards the middle proportionately as the temperature of the working steam approaches that of saturated steam. Even in the case of saturated steam, it is advisable to leave a neutral zone between the heating jacket and the cooling jacket, the exhaust belt being assumed to act as a cooling jacket. This neutral zone is neither heated nor cooled. It is also preferable to take the steam for the cylinder jacketting from the upper end of the cover where, in the case of superheated steam, the temperature is considerably reduced in consequence of the heat given up by the cover jacket. A further reduction in temperature takes place in this cylinder jacket, which of course does not exercise such a strong heating action as the cover jacket. The neutral zone,

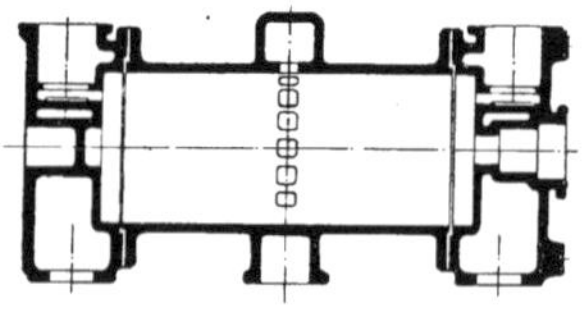

Fig. 11.

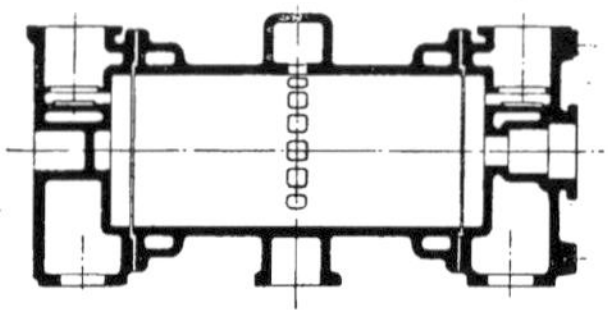

Fig. 12.

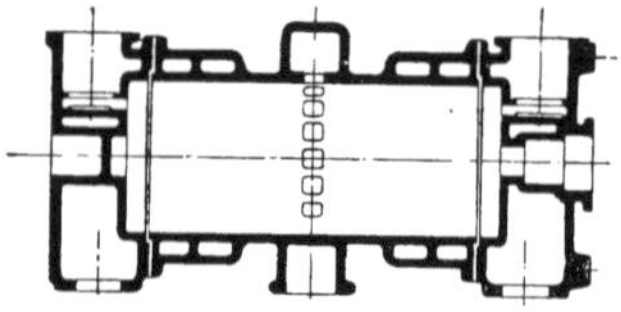

Fig. 13.

which is neither heated nor cooled, lies next to the moderately heated part, and then in the middle of the cylinder there is arranged the cool zone, represented by the exhaust steam belt, which is in the form of an annular casing around the central exhaust ports. In this way the principle of graduating the heating temperature is carried out in such a manner that the heating temperature, from the cover end towards the exhaust ports, decreases with the temperature of the working steam in the cylinder. The cooling jacket in the middle of the cylinder has the practical purpose of obtaining the best possible working conditions for the piston. At this point, where the piston velocity is greatest, the working temperature of the walls is lowest. The temperature of the cylinder wall is so fixed by the heating jackets that the surface temperature at the cover is greater than the temperature of the working medium, and in the middle of the cylinder the surface temperature is below the temperature of the working medium. The efficiency of the heating jacket of a steam cylinder, neglecting all unavoidable subsidiary losses, is given by the difference between the gross advantage of the heating and the loss in heat carried off from the cylinder by the exhaust. If the cover only is heated, a heating jacket is obtained in which, neglecting all unavoidable subsidiary losses, the gross gain is equal to the nett gain, that is to say the loss to the exhaust is nil. In the case of the cover jacket, it may be assumed that the entire heat from the cover is taken up by the steam layer adjacent to the cover. This quantity of steam corresponds very closely to the residual steam trapped in the cylinder immediately after the exhaust ports are closed by the piston on its return stroke. With a cover heating jacket working in such a manner, the steam which flows over the heated surfaces never passes out at the exhaust, so that no heat can ever be lost to the exhaust. The conditions in figure 12 are somewhat less favourable, and still less favourable are the conditions according to figure 13. In these two figures (12 and 13) the quantity of heat, which is carried off to the exhaust and is thereby lost, increases, because part of the steam which passes over the heated surfaces, goes off to exhaust. Experiments have shown that the culmination point, as regards advantage of jacketting, is passed, even for the case of saturated steam, when the heating jacket on the cylinder is extended right up to the exhaust belt. The jacketting is decidedly disadvantageous if it extends over the exhaust zone. The great loss incurred in such a case is easily explained by the great temperature difference and by the great velocity of flow of the exhaust. It is therefore preferable to have, in all cases, a neutral zone, the length of which should be determined according to the temperature of the working steam, that is to say, the neutral zone should be long in the case of highly superheated steam and short in the case of saturated steam.

Tests made with a cover-jacket, as shown in figure 11, showed a fall of 30° C, in the temperature of the superheated steam after passing through the jacket; that is to say, with steam entering at the inlet below with a temperature of 300° C, the temperature at the top, where the steam had completed its heating action in the cover, was 270° C, so that about 15 Centigrade heat units are given up to the working steam in the cylinder for each kilogram of steam used. If it is assumed that this quantity of heat is wholly taken up by the residual steam which is trapped in the cylinder after the piston closes the exhaust ports on its return stroke, or in other words, if it is assumed that none of the jacket heat is carried off by the exhaust, and assuming further that the rate of heat transference from the hot jacket to the working steam is directly proportional to the temperature difference between the jacket and the working fluid in the cylinder, figures 14 and 15 will represent on the entropy-temperature ($\Phi\tau$) diagram the thermic changes which take place. For each kilogram of saturated steam expanding from the initial pressure of 12 atmos.

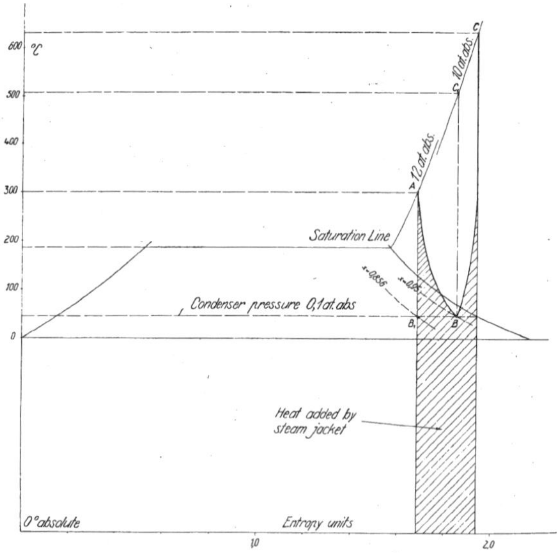

Fig. 14.

to the condenser pressure of 0·1 atmos., the residual steam trapped in the cylinder after exhaust will be 0·122 kg. With steam of the same initial and final pressures, but superheated to 300° C, the weight of the residual steam per kg. of working steam would be 0·142. In order to simplify the representation of the stages of the thermic changes in the diagram, the heat quantities have been worked out for the case of 1 kg of residual trapped steam. These quantities amount, in the case of saturated steam, to 82 C heat units, and in the case of superheated steam to 106 C heat units. With steam at 300° C, the heat transferred during expansion to the mass of steam, which will be trapped in the cylinder at the end of the next exhaust, is given by the line $A\ B$ from which it will be seen that there is a considerable increase in entropy. It is here assumed that the initial pressure is 12 atmos. and the final pressure 0·8 atmos. absolute in both cases, and also that the expansion continues during exhaust so that the point B corresponds to the condenser pressure of 0·1 atmos. The dryness fraction (x) of the residual steam is in the case taken 0·95 (B) instead of 0.856 (B_1). During the next compression stroke more heat is transferred to the steam, and the entropy is correspondingly increased. During this transference of heat the temperature of the compressed steam rises, so that the rate of trans-

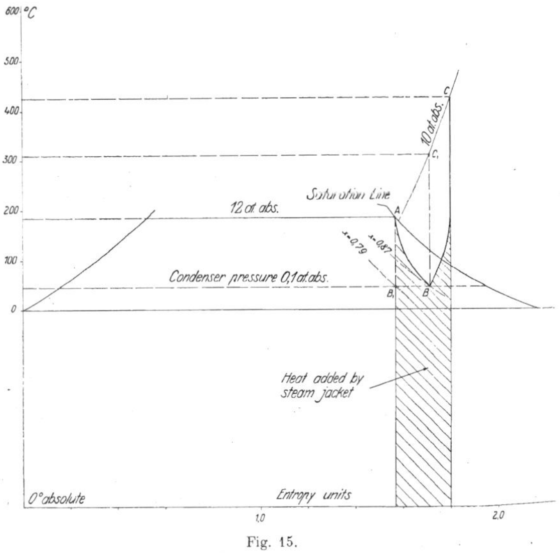

Fig. 15.

ference of heat from the jacket to the compressed steam decreases proportionally to the temperature difference. As soon as the compression temperature rises to the initial temperature (300° C) no further heat is transferred from the hot cover to the residual steam. The further compression is thus pure adiabatic, corresponding to the vertical line to C. The end temperature of compression at C is about 630° C. *The secret of the great efficiency of the una-flow engine resides, to a considerable extent, in this extraordinary rise in temperature.* Due to this great rise in temperature, the internal surfaces are most effectively heated to prepare them for the next charge of steam.

The high temperature at C also counteracts the conditions obtained at B. Thus although, practically, there is 5% moisture present at B, condensation or moistening of the cylinder walls at this point is prevented immediately by the heating action of the jacket and the adiabatic compression from B to C.

In figure 15, the same thermic changes are represented for the case of saturated steam at 12 atmos. initial pressure expanding down to 0·8 atmos. in a una-flow engine with a hot steam cover jacket. At the end of expansion, the steam which will be trapped (i. e. of course the layer nearest the hot jacket) has x (dryness fraction) $= 0 \cdot 87$ instead of 0·79, whilst the end temperature of the adiabatic compression is about 425° C, at the point C. It will be seen therefore that even in the case of saturated steam, there is a very considerable heating of the internal

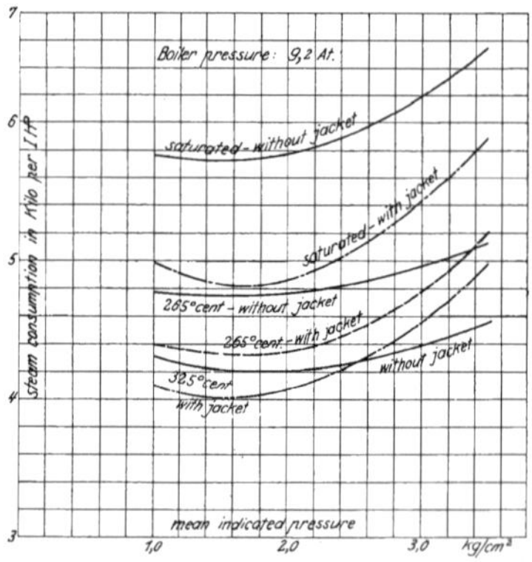

Fig. 16.

surfaces of the cylinder by the compressed steam. The heating action is so considerable that, even in the case of saturated steam, cylinder condensation or moistening of the walls or clearance surfaces is scarcely possible.

The shaded parts of each of the figures 14 and 15 represent the amount of heat transferred from the hot jacket to the working steam, this amount being 82 Centigrade heat units in the case of saturated steam (fig. 15) and 106 Centigrade heat units in the case of superheated steam at 300° C (fig. 14). Although other conditions present may influence the thermic changes which have been described, figures 14 and 15 nevertheless give some idea of the chief outstanding thermic changes, as is confirmed by tests made by Prof. Hubert and M. A. Duchesne described more fully later with reference to figures 33—38.

The above considerations and deductions are in part confirmed by the steam consumption tests of a una-flow engine shown graphically in figure 16. These tests were made with a una-flow steam engine provided with a cover-jacket similar to that which is made the basis of the above investigations. In addition, jackets were arranged over the cylinder at each end, and the centre part of the cylinder was cooled by the exhaust belt. Between the exhaust belt and the cylinder jackets there were neutral zones which were neither heated nor cooled. The cylinder jackets were arranged so that they could be cut out. During the experiments the cover jacket was *always* in use, but the cylinder jackets were sometimes put in and sometimes cut out. The results with the cylinder jackets in use are marked "with jacket"; those where the cylinder jackets were cut out are marked "without jacket". First of all, the results show that the advantage of jacketting is less as the temperature increases. In the case of saturated steam, the surprising difference of 1 kg. in favour of jacketting is obtained, whilst the saving is barely $\frac{1}{2}$ kg. in the case of steam superheated to 265° C, and $^{2}/_{10}$ ths kg. in the case of steam superheated to 325° C. All the above figures are taken at the points where the steam consumption per H.P. was lowest.

For the case of the most economicl cut-off, (considering also the first cost of the engine) with steam superheated to 325° C, there is no difference in the steam consumption per I. H.P. with and without cylinder jacketting. In all cases the steam used in the jacket was taken into account and added to the steam used in the cylinder.

With a mean pressure of 2·5 kgs. per □ cm, the consumption per I. H.P. with cylinder jacketting was the same as the steam consumption per I. H.P. without jacketting, in both cases the steam being superheated to 325° C. With steam at 265° C the corresponding mean pressure is about 3·4 atmos., that is the mean pressure at the point of intersection of the two curves. The point of intersection of the two curves in the case of saturated steam is still further to the right in figure 16, and corresponds to a still higher mean pressure.

On comparing the steam consumption curves for a una-flow engine with those for a multiple expansion steam engine, it is found that in the case of the una-flow engine the steam consumption does not vary so much with the load. This will be noticed chiefly in the case of the curves without jacketting, where no material variation occurs in the steam consumption between 1 and 3 atmos. mean pressure when the superheat is high. In the curve for saturated steam, little difference in the steam consumption occurs between 1 and 2·4 atmos. when a jacket is provided on the cylinder.

It is to be noted further, that with steam at the moderate pressure of 9·2 atmos. and superheated to 325° C the lowest steam consumption is very nearly 4 kgs. per I. H.P. *This is to say, in the case of a una-flow engine of scarcely 300 H.P. the same steam consumption was obtained as was attained by the 6000 H.P. triple expansion engines of the Berlin Electricitätswerk in Moabit. According to the figures published by Herr Datterer the lowest steam consumption obtained by these engines was 4 to 4·1 kgs. per H.P.*

The corresponding consumptions for the cases taken in figure 16, are represented in figure 17 in heat units. This figure shows that at the most economical cut-off the heat consumption is the same, with moderate superheat to 265° C and cylinder jacketting, as with high superheat at 325° C either with or without jacketting.

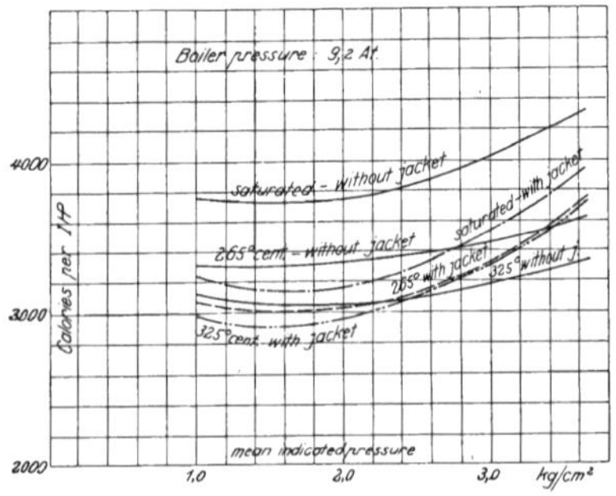

Fig. 17.

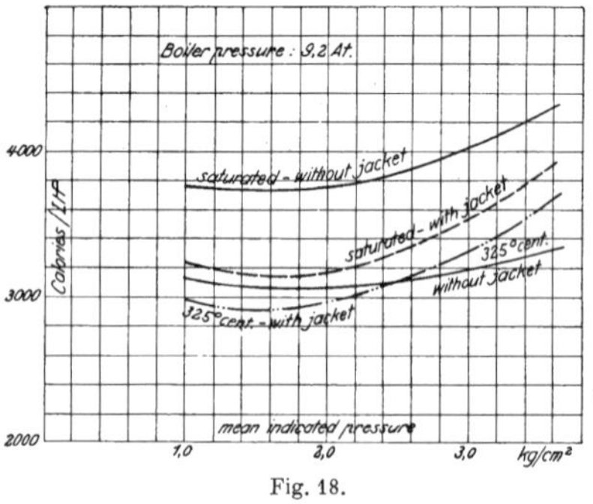

Fig. 18.

In figure 18, the heat curves corresponding to figure 17, are repeated, but the curves for moderate superheat to 265° C, have been omitted. In this case, as in figure 17, all recovery of heat from the water of condensation passing from the jacketting has been neglected. It is nevertheless evident from figure 18, that the minimum heat consumption, in the case of saturated steam working with the cylinder jacket, approximates very closely to the lowest heat consumption in the case of highly superheated steam working without any jacket on the cylinder walls.

In figure 19, the heat curves given in figure 18, are shown as corrected for the heat recovered from the water of condensation passing from the jacket. This figure shows that, when the curves are corrected in this way, the lowest heat consumptions for saturated steam with jacketting, and for superheated steam at 325° C without jacketting, are practically the same in the same engine.

Taking the Berlin conditions, viz, that 1 kg. steam corresponding to 725 Centigrade heat units costs 0·225 pfennigs, and that for superheated steam $^2/_3$ gr of oil per I. H.P. hour is required = 0·05 pfg. (taking oil at 75 marks per 100 kgs.) whilst for saturated steam only $^1/_3$ gr of oil per I. H.P. hour = 0·02 pfgs. (taking oil at 60 marks per 100 kgs.) is required, and calculating out these oil charges in heat units (160 centigrade heat units for superheated and 64 centigrade heat units for saturated steam) the results shown in figure 20 are obtained. It will be seen from figure 20 that the steam and lubrication costs taken together are almost the same at the most economical cut-off in both cases, that is in the case of saturated steam working with cylinder jacketting, and superheated steam without cylinder jacketting. In both cases of course there is cover-jacketting. *The above clearly shows that with a una-flow engine it is almost immaterial, as regards economy, whether the steam is superheated or not. It should be pointed out, however, that in this case the "saturated" steam was superheated a few degrees to ensure its dryness.*

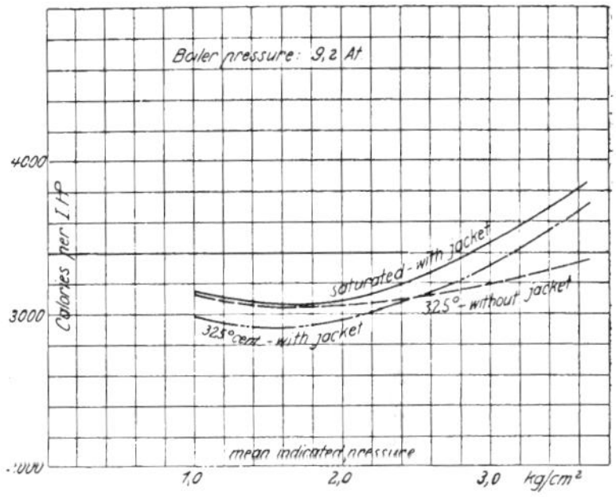

Fig. 19.

In the above investigations no account has been taken of the fact that heat radiation and leakage losses in the pipes and steam distributing parts are much higher in the case of superheated steam than in the case of saturated steam. This increase in these losses is due to the decreased density of the superheated steam and to theg reater strain on the working parts. In the engine investigated, all the distributing parts were tight in both cases.

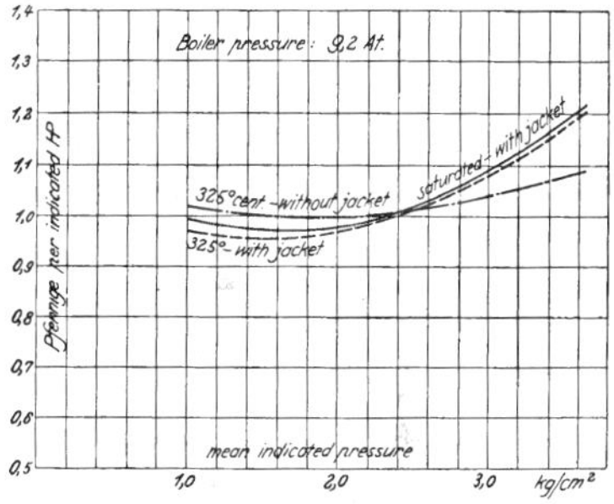

Fig. 20.

It may also be seen from consideration of the character of the adiabatic expansion and compression curves, that the una-flow steam engine must give almost the same economic efficiency with superheated as with saturated steam. This fact leads one to conclude that there is such absence of heat losses that the extraction of energy from the steam is almost exactly represented on the corresponding entropy temperature diagram. A comparison of the entropy temperature diagram for saturated and superheated steam, between the same pressure limits, shows only a very small advantage in favour of superheating. This small advantage may be, however, entirely removed if the heat in the water of condensation, passing from the heating jacket, in the case of the saturated steam cycle, is returned to the boiler to be recovered therein for useful purposes. The great mistake in steam engines of the ordinary construction resides in the counter-flow action. By reversing the flow of the steam there is always considerable cooling of the clearance surfaces by the cold exhaust current, which places these surfaces in the worst possible condition for receiving the next charge. The consequence is, necessarily, considerable initial cylinder condensation which, in the case of the ordinary counter-flow engine, is sought to be avoided by extracting the energy in stages and by superheating. Stages and superheating are therefore "make shifts" used to correct the injurious results of the principal error in the ordinary form of steam engine construction which, throughout this work, is referred to as the counter-flow type.

If now these fundamental mistakes are avoided, that is to say, if the counter-flow is replaced by a uni-directional flow, the cooling of the clearance surfaces is avoided so that the extraction of the energy of the steam in stages and superheating become superfluous.

This result, which has been demonstrated by ordinary common sense reasoning, is confirmed entirely by the deductions made from the above tests.

The above results were obtained in an engine which might be said to be of fairly good design and excellent workmanship. It was the first una-flow engine built by Gebr. Sulzer after a design by the author.

Figures 21 and 22, show the efficiencies obtained in the tests with steam superheated to 325° C, and with saturated steam. The efficiencies given take into account all losses due to throttling and leakage as well as thermal losses. The efficiencies are calculated as the ratio of the mean pressures, obtained at various cut-offs in the actual engine, to the corresponding mean pressures in the ideal engine. In the ideal engine, the thermal losses, as well as the losses due to throttling and leakage, were assumed to be nil. The efficiency in both cases reaches the high value of 95%, that is to say the losses mentioned total together only 5%, a result which could scarcely be surpassed.

The more favourable utilisation of the heat in the case of superheated steam has not been considered in the above comparisons, this being more clearly seen in the entropy-temperature diagram. It is to be remembered however, that in the case of superheating, the increased lubrication costs, the increased radiation losses in the pipe lines, and frequently a lower boiler efficiency, have to be written down on the debit side. Consideration should also be given to the fact that leakage losses are frequently increased by the use of superheated steam.

Figures 23—26 show the constructions designed to satisfy the conditions and to give the results set out above. It is always advisable to heat the cover. It is also advisable to provide a jacket over the cylinder at its ends with steam temperatures up to 325° C. It is further advisable to divide this jacket into compartments which, in the case of small cylinders, may be two in number; but in large cylinders three or more subdivisions are advisable.

The hot steam should be led into the end section or compartment of the cylinder jacket, and should be led from the first into the second compartment at the diametrically opposite point, and then again from the second to the third compartment at the diametrically opposite side, so that the steam passes in sequence and in a zig-zag path through the various compartments, commencing at the compartment nearest to the cover. In this way there is a graduation of the temperature, especially with superheated steam, which commences with its maximum

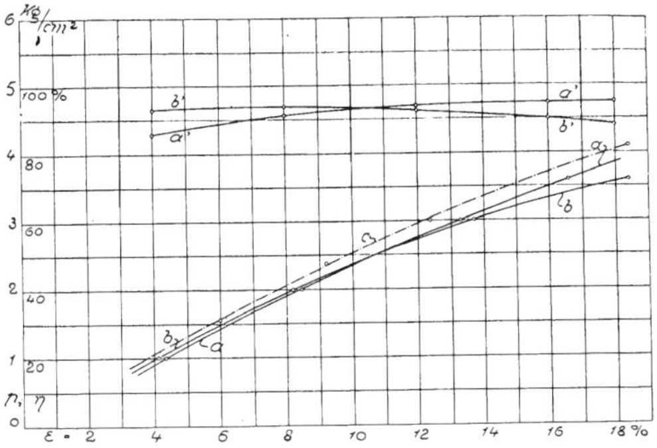

Fig. 21.

Efficiencies of a 300 H.P. una flow engine working with superheated steam (325° C).

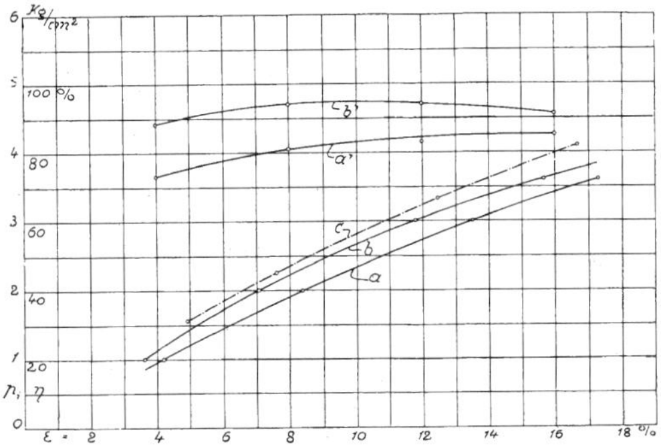

Fig. 22.

Efficiencies of a 300 H.P. una-flow engine working with saturated steam.

In figures 21 and 22

line *a* shows the actual mean pressures without jacket,
,, *b* ,, the actual mean pressures with jacket,
,, *c* ,, the mean pressures of the ideal engine,
,, *a'* ,, the efficiencies without jacket, ratio *a*:*c*,
,, *b'* ,, the efficiencies with jacket, ratio *b*:*c*,

the base line representing the steam consumption per stroke in percentages of the cylinder volume.

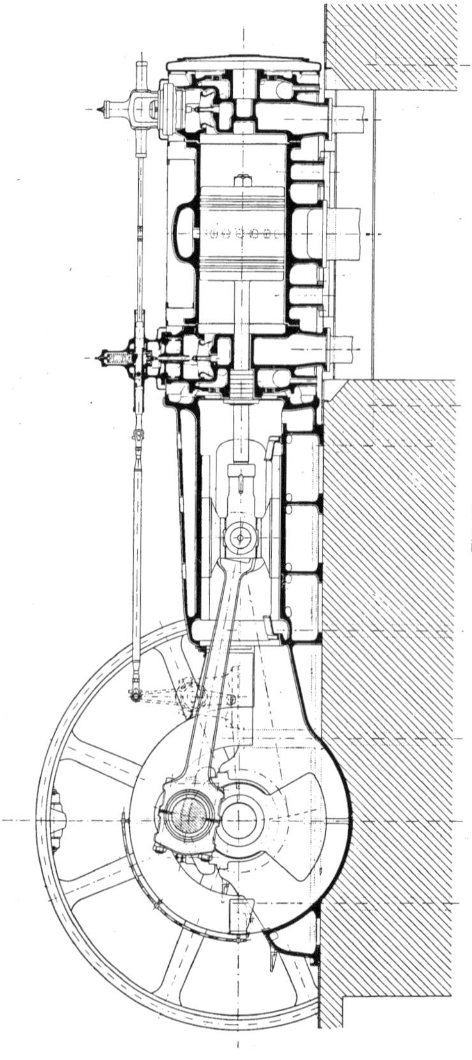

Fig. 25.

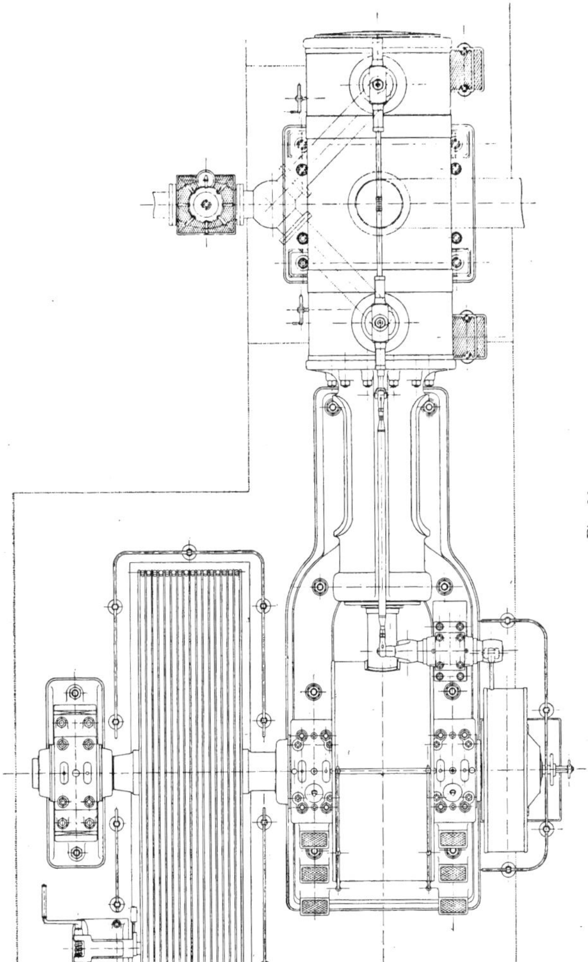

Fig. 26.

at the inlet end and decreases towards the exhaust, being limited at the hot end by the hot cover jacket and at the cold end by the neutral zone and the exhaust belt.

The action of the above constructional features is seen in an interesting light by referring back to figure 4, where the cylinder wall temperatures are plotted on the same diagram as the saturated steam temperatures at various points in the

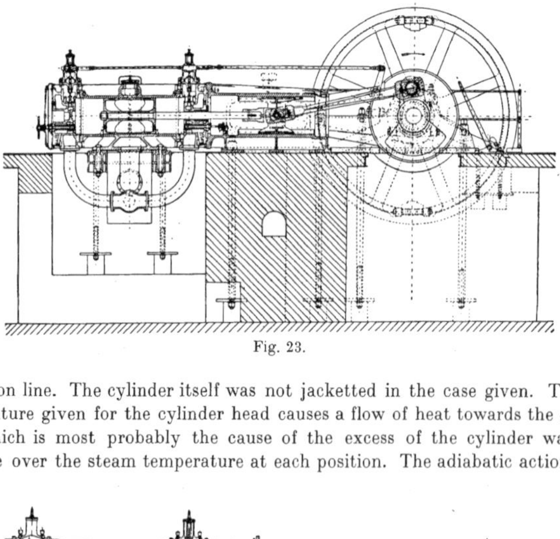

Fig. 23.

expansion line. The cylinder itself was not jacketted in the case given. The high temperature given for the cylinder head causes a flow of heat towards the exhaust belt, which is most probably the cause of the excess of the cylinder wall temperature over the steam temperature at each position. The adiabatic action of the

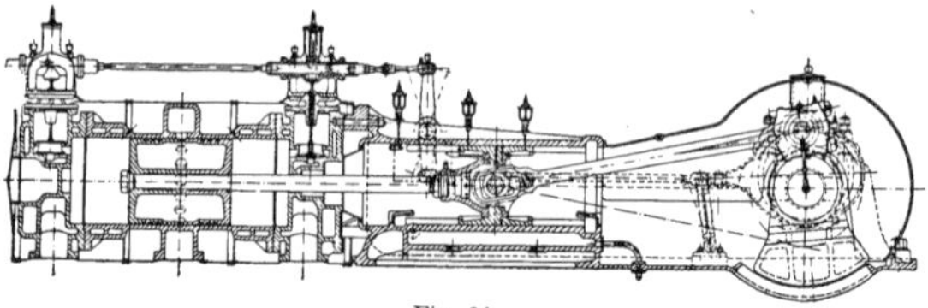

Fig. 24.

cylinder walls may also be traced to this cause and the great heat saving may be thus explained. It would appear therefore that it is of the utmost importance for the cylinder wall temperature at the ends to be in excess of the working steam temperature, and at the middle, to be below the exhaust temperature. This was confirmed by measurement of the cylinder wall temperatures and the steam consumption obtained. The actual tests completely bear out this explanation with different temperatures of steam and different jacketting.

Chapter IV.
The prevention of leakage.

Absolutely steam tight valves and steam distributing parts are very exceptional in steam engines generally. A flat slide valve is usually considered more steam-tight than a piston valve, for which reason many firms adopt the plan of employing a flat slide valve for the L.P. cylinder and a piston valve for the H.P. cylinder. This rule is to be recommended for meeting the conditions present in the two cylinders as regards pressure and temperature. Corliss valves, to a considerable extent, are steamtight, but even in the case of Corliss valves, it cannot be said that *absolute* steam tightness is obtained.

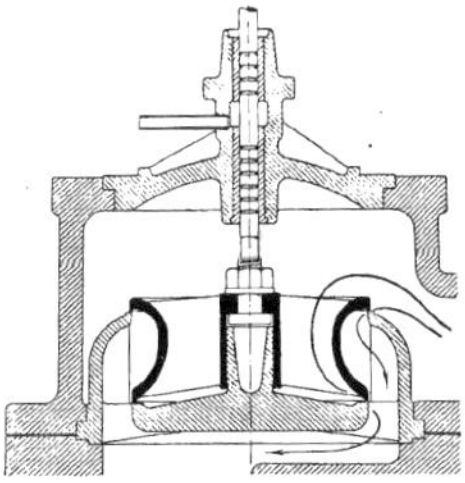

Fig. 27.

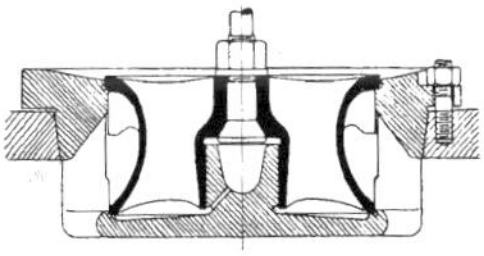

Fig. 28.

Piston valves may be made steam tight if the workmanship is first class and if spring pressed packing rings are employed. Piston valves without springs may be made fairly steam tight in very small sizes. With larger sizes and superheated steam, piston valves should always be provided with spring packing rings, if any reasonable degree of steam-tightness is required.

In most cases, balanced lift valves have some leakage. The amount of the leakage increases with the size of the valve, the degree of the balance, the pressure and the temperature. In all forms of valves, the amount of leakage is increased by superheating, which causes distortions of the parts whilst at the same time the steam is less dense.

Both pressure and temperature have a bad influence on the tightness of valves. In a double beat valve, as shown in figs. 27, 28, during expansion there is a pressure on the valve in the direction of its axis, and the valve is correspondingly compressed so as to raise the bottom valve face from its seat and cause leakage. At the same time the lower seat of the casing is pressed down by the steam pressure acting on the lower circular area of the casing. In the design according to fig. 28 the ribs connecting the lower plate of the casing with the upper ring will be lengthened, thereby causing additional leakage. The leakage will be increased in proportion to the steam pressure, the reduction of diamater of the waisted part of the valve and the length and diameter of valve and casing. The radial forces on the valve may be neglected when the valve seats are flat. At high temperatures, the valve expands more than the seat and the casing. The consequence of this is, that the valve presses hard on its lower seat and rises slightly from its upper seat, thereby causing leakage at the upper valve face. Unequal temperatures may be caused in different ways. In the construction according to fig. 27, in which the valve and the seat have the same height and thickness, the expansion of the valve and its seat will only be equal if the expansion co-efficients of the metals used in both cases are equal. If care is taken during the casting, this result may be obtained. Both the parts are subjected on one side to the same live steam temperature, and on the other side to the same temperature of the expanded steam.

The conditions are not so favourable in the case of the valve shown in fig. 28, in which the valve face is of the ordinary construction. In this valve, the parts are of unequal thickness and therefore they expand unequally, at any rate in the time immediately following starting. During the whole time the engine is running, the valve is subjected on the one side to the temperature of the live steam, and on the other side to the temperature of the expanded or working steam in the cylinder. The ribs of the valve seats, on the other

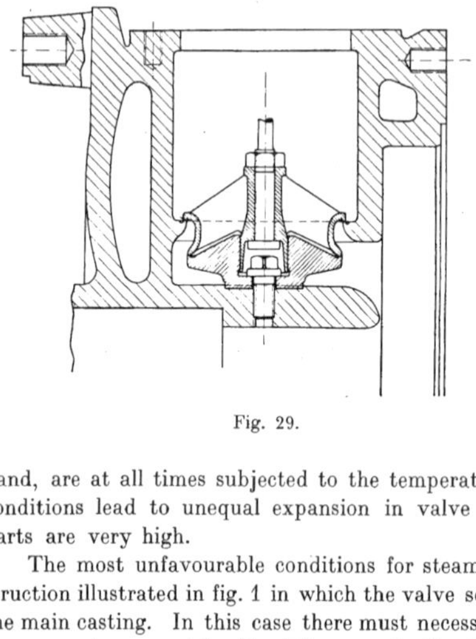

Fig. 29.

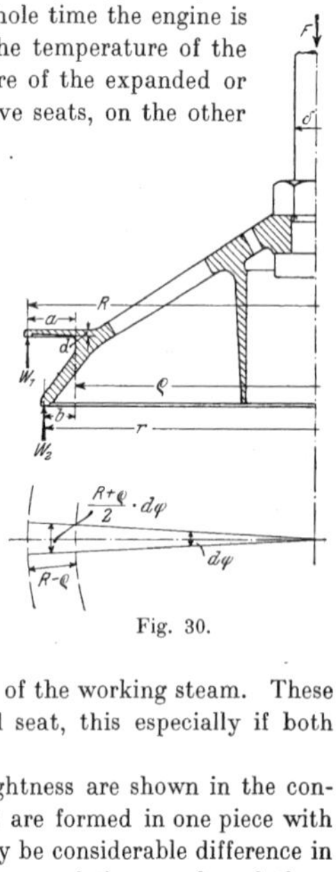

Fig. 30.

hand, are at all times subjected to the temperature of the working steam. These conditions lead to unequal expansion in valve and seat, this especially if both parts are very high.

The most unfavourable conditions for steam tightness are shown in the construction illustrated in fig. 1 in which the valve seats are formed in one piece with the main casting. In this case there must necessarily be considerable difference in the expansion caused by the differences in the thickness of the metal, and there must also be a corresponding amount of leakage, especially immediately after starting. In all cases and under all circumstances, inequalities in the expansion coefficients and temperature differences are present. It follows from this that the construction according to fig. 1, will not be tight.

The leakage always increases with increase in the height of the valve. It is therefore advisable to make the valve as short as possible. For this reason all valve gears which require an unnecessarily large lift of the valve and consequently involve the use of high valves (if the upper valve opening is not to be throttled), should be avoided. It is preferable on this account to employ a roller or cam lift for the valve, by means of which the valve is merely raised the necessary amount. It is also advisable, to employ a chair which carries the lower valve seat (see fig. 29), especially when a valve construction with seat cases such as shown in figure 28, and similar forms, are to be avoided for the purpose of reducing clearance surfaces and the number of packings. In this form of valve the amount of the relative expansion between the valve and valve seats is reduced owing to the reduction of the height of the valve. The small reduction of the diameter of the valve at the waisted part renders the valve much more capable of resisting bending when vertical forces are applied. With a construction of this kind it is possible to get a very close approximation to absolute steam tightness up to a fixed pressure and temperature if the valve is ground in place at the working temperature. Immediately after starting and with varying pressures and temperatures a certain amount of leakage occurs even in this form. It is remotely possible at fixed temperatures and pressures and by employing a conical seat to obtain complete steam tightness. *Absolute steam tightness under varying conditions of pressure and temperature would appear to be obtainable only with a resilient valve of a construction, such as shown in figure 30.*

The lower seat is of smaller diameter than the upper seat. In this way a certain amount of steam pressure (i. e. the steam pressure difference on the overhanging surface) is used for closing the valve in addition to the spring pressure. The resilient seat permits of inequalities in the expansion of the parts. Even in this case it is preferable to arrange the lower valve seat on a chair in order to keep the valve as short as possible and reduce the amount of the difference in expansion between the valve and its casing.

Calculations for the resilient valve.

The following forces act on the valve.

Down.

1. The spring pressure, less the steam pressure on the spindle.

$$P = F - \frac{\delta^2 \pi}{4} (p_a - 1);$$

p_a = absolute pressure in valve case.

2. The pressure on the upper side of the annulus

$$(R^2 - \varrho^2)\, \pi\, (p_a - p_i);$$

p_i = absolute pressure within the cylinder.

Upwards.

3. The pressure on the lower side of the lower annulus

$$(r^2 - \varrho^2)\, \pi\, (p_a - p_i).$$

4. The upward bearing pressure of the upper valve seat (W_1).
5. The upward bearing pressure of the lower valve seat (W_2).

The horizontal forces balance each other.

The sum of the vertical forces must be nil so that:

$$P + (R^2 - \varrho^2)\, \pi\, (p_a - p_i) - (r^2 - \varrho^2)\, \pi\, (p_a - p_i) - W_1 - W_2 = 0 \qquad (1)$$

If the radii, the steam pressures and the spring pressure are known, only the bearing pressures W_1 and W_2 remain as unknown. W_1 may be estimated from the condition that the bending (f_1) of the resilient ring (measured in the seat) due to the steam pressure must be balanced by the bending (f_2), due to the upward pressure W_1 of the upper seat, that is if the lower valve face is to remain on its seat.

This is to say

$$f_1 = f_2.$$

Imagine a piece cut out radially from the valve with the angle $d\varphi$ at the centre, then the steam pressure acting on this section of the resilient surface is

$$\left(\frac{R+\varrho}{2}\right) \cdot d\varphi\,(R-\varrho)\cdot(p_a - p_i).$$

and the bending caused by this pressure is

$$f_1 = \left(\frac{R+\varrho}{2}\right) \cdot d\varphi \cdot (R-\varrho)\,(p_a - p_i)\,\frac{(R-\varrho)^3}{8\,E\,J} \quad . \quad . \quad . \quad . \quad (2)$$

Where E = modulus of elasticity and J = the moment of inertia of the cross section at right angles to the plane (approximately constant) of bending.

The upward pressure (W_1) of the upper seat acting on the radial section under consideration is

$$W_1 \frac{d\varphi}{2\pi}$$

and the consequent bending is

$$f_2 = W_1 \frac{d\varphi}{2\pi} \cdot \frac{(R-\varrho)^3}{3\,E \cdot J} \quad . \quad . \quad . \quad . \quad . \quad . \quad . \quad . \quad . \quad (3)$$

Then on equating $f_1 = f_2$.

$$W_1 = \frac{3}{8}\pi\,(R+\varrho)\,(R-\varrho)\,(p_a - p_i) \quad . \quad . \quad . \quad . \quad . \quad (4)$$

From equation (1)

$$W_2 = P + (p_a - p_i)\,(R^2\pi - r^2\pi) - W_1 \quad . \quad . \quad . \quad . \quad . \quad (5)$$

Substitute in (5) the value found in (4) for W_1. Then

$$W_2 = P + \left(\frac{5}{8}R^2 - r^2 + \frac{3}{8}\varrho^2\right)\pi\,(p_a - p_i) \quad . \quad . \quad . \quad . \quad . \quad (6)$$

In the limit when the valve just rests on the lower seat, $W_2 = 0$. Neglecting P, the excess of the spring pressure over the spindle pressure, the highest permissible value for ϱ may be obtained as follows

$$\frac{5}{8}R^2 - r^2 + \frac{3}{8}\varrho^2 = 0 \quad . \quad . \quad . \quad . \quad . \quad . \quad . \quad . \quad . \quad . \quad (7)$$

Denote R by $\varrho + a$, and r by $\varrho + b$, the equation (7) becomes

$$\frac{5}{8}(\varrho + a)^2 - (\varrho + b)^2 + \frac{3}{8}\varrho^2 = 0$$

$$\frac{10}{8}\varrho \cdot a + \frac{5}{8}a^2 - 2\varrho b - b^2 = 0$$

$$\frac{5}{8}a\,(a + 2\varrho) - b\,(b + 2\varrho) = 0$$

or since $(a + 2\varrho)$ and $(b + 2\varrho)$ are approximately equal

$$b = \frac{5}{8} a \quad \ldots\ldots\ldots\ldots \quad (8)$$

If the valve expands by the amount Δl in excess of the casing, in consequence of unequal temperatures, the valve must bend by the amount Δl at the resilient seat in order to obtain steam tightness. In this case $f_1 - f_2$ is not nil but is equal to

$$\Delta l = f_1 - f_2 \quad \ldots\ldots\ldots\ldots \quad (9)$$

The bearing pressure W_1 for any given Δl is obtained by setting the values for f_1 (equation 2) and f_2 (equation 3) in equation (9); that is to say:

$$W_1 = \frac{\pi}{2}\left[\frac{3}{4}(R^2 - \varrho^2)(p_a - p_i) - \Delta l \cdot E \cdot \varrho \left(\frac{d}{R - \varrho}\right)^3\right] \quad \ldots \quad (10)$$

For the valve to be tight, W_1 must be positive, for which purpose a pressure difference $(p_a - p_i)$ is necessary, as may be seen from equation (10). The lowest pressure difference required for tightness may be obtained from equation (10) by making $W_1 = 0$, i. e.

$$p_a - p_i = \frac{4 \Delta l \cdot E \cdot \varrho}{3 (R^2 - \varrho^2)} \left(\frac{d}{R - \varrho}\right)^3 \quad \ldots\ldots\ldots \quad (11)$$

From this equation (11) it will be seen that the pressure difference is proportional to Δl and consequently to l. To retain $(p_a - p_i)$ small, the valve must be short. The value given to d, is determined by the strength of the material and the values given to R and ϱ are determined by the desired velocity of flow of the steam past the valve.

To determine the bending stresses on the resilient ring during working:

Let t_1 = denote the temperature of the valve,
,, a_1 = the expansion coefficient of the valve,
,, t_2 = the temperature of the surrounding casing,
,, a_2 = the expansion coefficient of the casing,
,, l = the distance apart of the two seats at the normal temperature t_0.

Then

$$\Delta l = l\,[a_1 (t_1 - t_0) - a_2 (t_2 - t_0)] \quad \ldots\ldots\ldots \quad (12)$$

Substituting this value in equation (11) the pressure difference $(p_a - p_i)$ will be found at which the valve will commence to be tight. At all greater pressures the valve will be quite tight. Assume a valve in which $l = 30$ mm, $R = 125$ mm, $\varrho = 104$ mm, $d = 3$ mm, $p_a = 12$ atmos. and $p_i = 0$ atmos. Then take

$$t_0 = 15^0\ \text{C},$$
$$t_1 = 300^0\ \text{C},$$
$$a_1 = 0{\cdot}000012,$$
$$a_2 = 0{\cdot}000011.$$

In the following table the relative expansions of this valve for certain temperature differences $(t_1 - t_2)$ calculated according to equation (12) are given. The stress K_b and the pressure differences $(p_a - p_i)$ according to equation 11 are also given.

$t_1 - t_2$ =	0°	50°	100°	150°	200°
Δl = in mm	0·009	0·0255	0·042	0·0583	0·075
$p_a - p_i$ = in atm. abs.	1·35	3·8	6·3	8·75	11·25
K_b =	230	645	1070	1490	1920

From the above figures it will be seen that many prevalent constructions of valves are far from steam tight. The excessive length of many valves, neglecting other constructional errors, makes steam tightness absolutely impossible. Even in a valve of only 30 mm in length with a temperature difference of 200° C between the valve and its seat, a steam pressure of 11·25 atmos. is necessary in order to get steam tightness. That is to say, at all lower pressures there is leakage.

Calculations for a resilient valve, 160 mm mean diameter for a condensing engine working with steam at 9 atmos. absolute (fig. 31).

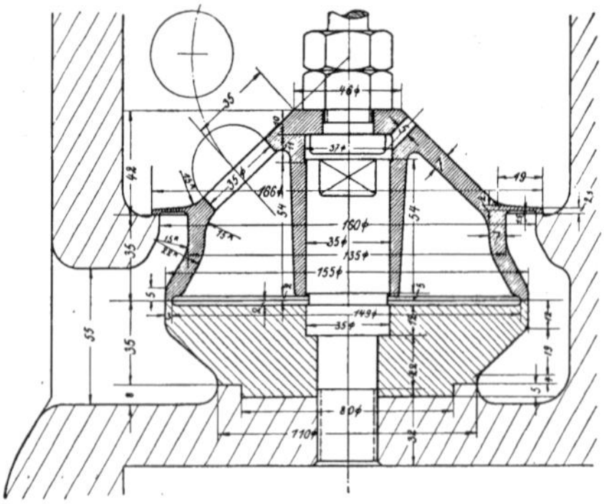

Fig. 31.

The following two conditions are assumed for the calculations of W_1 and W_2:

(I) the actual closure occurs at the mid line in both the top and bottom faces (Radii R_m and r_m);

(II) in the most unfavourable conditions, the closure is at the inner edge of the seat (R_i) in the upper valve face and at the outer edge (r_a) in the lower face.

In the latter case W_2 on the lower valve seat should be 0.

In figure 31

$$R_m = 81{\cdot}5 \text{ mm}; \quad R_i = 80 \text{ mm}; \quad \varrho = 67{\cdot}5 \text{ mm},$$
$$r_m = 76 \text{ mm}; \quad r_a = 77{\cdot}5 \text{ mm}; \quad \delta = 17 \text{ mm}.$$

The spindle pressure

$$\frac{\delta^2 \pi}{4} (p_a - 1) = 18 \text{ kgs.}$$

W_1 is, according to equation (4) above,

for R_m and r_m = 221 kgs,
for R_i and r_a = 195.5 kgs.

$W_2 = 0$ in case (II). The spring pressure (F) is then calculated from equation (5).

$$F = \frac{\delta^2 \pi}{4} (p_a - 1) + W_1 - (p_a - p_i)\ (R_i^2 \pi - r_a^2 \pi) = 18 + 195{\cdot}5 - 107.$$

$$F = 106{\cdot}5 \text{ kgs.}$$

With this spring pressure in case (I)

$$W_2 = P + (p_a - p_i)\ (R_m^2 \pi - r_m^2 \pi) - W_1 = 88{\cdot}5 + 245 - 221,$$
$$W_2 = 112.5 \text{ kgs.}$$

The thickness d, of the resilient ring is calculated for the case of the valve being opened at the greatest pressure difference $p_a - p_i$; the valve and seat being assumed to be at the same temperature.

Take once again a thin radial strip $d\varphi$ (fig. 30) of the valve then is

$$M_b = (R - \varrho) \frac{R + \varrho}{2} d\varphi\ (p_a - p_i) \frac{R - \varrho}{2} = \frac{1}{6} \varrho \cdot d\varphi \cdot d^2 \cdot K_b$$

and

$$K_b = \frac{3}{2} \frac{(R + \varrho)\ (R - \varrho)^2}{\varrho\, d^2} (p_a - p_i).$$

If d is chosen at 2 mm then for the value R_m will be

$$K_b = 1460 \text{ kg per cm}^2.$$

These high stresses only occur when the full pressure is on the valve and no back pressure in the cylinder — a case which scarcely ever arises.

The calculations show the advisability of making the valve of forged steel (fig. 31), as with this material, the resilient ring or tongue may be made very thin and its radial width may be very small for any reasonable degree of flexion.

From the above calculations it follows that the breadth of the unbalanced ring, measured from the mid point on one seat to the mid point of the other, is $^3/_8$ the radial width of the resilient ring or tongue, and this will give the minimum loading of the valve and the valve gear. In conjunction with this, the governing of the engine may be improved, as a governor of smaller power may be employed.

By means of a resilient valve of forged steel, used in conjunction with a chair for the bottom seat, which enables the valve to be of minimum height, the seat case used for many valves may be omitted and the clearance surfaces reduced, whilst obtaining perfect steam tightness under all conditions of working. It was shown on placing a burning candle in front of the open indicator cock that no disturbances or flickering of the flame occured although the valve was being subjected to a pressure of 12 atms.

With a construction such as described it is possible in the smaller sizes to obtain perfect steam tightness even with widely varying conditions of temperature and pressure.

If the unbalanced ring of the valve is smaller than found by the above theory, the lower valve face rises from its seat and causes considerable leakage. It is therefore always advisable to make a second calculation for fixing the bearing pressure on the lower seat, and in this calculation the inner diameter of the upper seat and the outer diameter of the lower seat should be employed. With these most unfavourable conditions the value of the upward bearing pressure of the lower seat should in this case be nil, assuming the valve to be still close up to its seats.

The valve seats should, under all circumstances, be flat, because with a flat seat, inequalities in radial expansions meet with no opposition or resistance, whilst at the same time the maximum amount of free valve opening is obtained. This latter feature is of the utmost importance, in view of the early cut-offs employed in the una-flow engine.

To prevent the lip of the valve from warping, a circular rib should be placed over the upper valve seat.

It will be understood that the resilient valve described has its limits, as regards its adaptability to a valve seat with changing situation. In smaller engines it is possible to form the seat in one with the cover casting or to carry the lower seat only on a separate chair fixed to the casting. In larger sizes it is advisable to connect the lower part of the chair with the upper by cast on radial ribs, because, when using higher superheats, internal strains introduced during casting might be set up, thereby causing considerable distortion. The surest construction is to connect the upper seat to the lower by ribs, so that the valve seats are carried by a basket, and their relation to one another does not vary. *This basket construction, in conjunction with a properly designed and manufactured resilient valve, will prove the best possible means for obtaining effective steam-tightness.*

Instead of using a resilient valve, the necessary steam-tightness may be secured by means of a low rigid valve and resilient chair for the bottom seat. This construction is shown in figure 32. The chair is of steel and has a corrugated or bellied annulus which supports the seat. The valve itself in this case may be made of cast-iron. The upper, rigid, seat is of larger diameter than the lower seat. When the valve is closed the resilient annulus should not be deflected. The annulus is designed so that the steam pressure will cause the lower seat to press upwards with a specific pressure equal or approximately equal to the specific pressure on the upper seat. Calling the spring pressure F and the steam pressure S, with a valve having the upper seat double the area of the lower seat the pressures would be $^2/_3\ (S+F)$ on the upper seat and $^1/_3\ (S+F)$ on the lower seat. In this case the overhang of the annular lip or bellied part would be dimensioned to give $^1/_3\ (S+F)$ upward pressure against the valve seat or $^2/_3\ (S+F)$ total pressure, this total pressure to be taken up on one side by the connection with the chair, on the other side by the valve. The pressure acting on the lip is to be calculated for the particular boiler pressure with which it has to be used. The resilient chair should also be arranged to compensate for differences in expansion between the valve and its casing.

In the una-flow engine the piston packing is arranged between the inlet valve and the exhaust. These three, namely, the inlet valve, the piston and the exhaust are arranged in series, whilst in the ordinary engine the piston is arranged outside the other two, that is to say the packings, for the inlet and the exhaust valves are

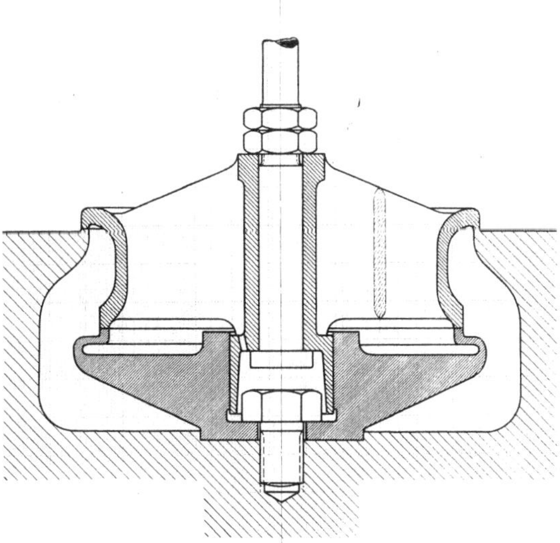

Fig. 32.

arranged to one and the same side of the piston. An advantage of no small importance for the una-flow engine lies in this arrangement.

In the counter-flow steam engine, leakage of the inlet valve is compensated for by leakage of the exhaust valve, and only the difference between the two leakages has an effect upon and produces changes in the contents of the cylinder. In the una-flow engine, however, there is no exhaust valve to compensate for leakage of the inlet valve. The exhaust valve in the case of the counter-flow steam engine remains open till shortly before the end of the stroke so that the simultaneous occurrence and probable compensating action of the two leakages is only effective during the short time of the compression. Turning now to the una-flow engine, any leakage of the inlet valve during the entire period of compression, that is to say during 90% of the stroke, has an effect on the contents of the cylinder. This leakage steam will fill up the clearance space. It will of course represent a loss, but not to the extent present in a counter-flow engine. In the latter case the leakage steam represents a total loss, whilst in the una-flow engine it will do useful work during expansion and only the additional work in compression will represent the loss. On the other hand, the engine is subjected to a check or jerk when the compression pressure exceeds the inlet pressure.

The construction of a really good and practical una-flow engine therefore rendered it absolutely necessary to devise a thoroughly steam tight inlet valve, and it was arising from this imperative necessity that the above constructions of resilient valve or resilient seat were evolved. The quick rise of the compression line, frequently observed in ordinary engines, may be explained by the compression steam taking up heat from the admission steam, and also by the addition of steam which has leaked past the inlet valve and been superheated during the throttling thereby involved, whilst at the same time, wet steam passes from the cylinder by leakage through the exhaust valve.

Chapter V.

The loss by the clearance surface.

In devising the una-flow engine the author has made an attempt to approximate to the conditions obtained in a steam turbine. As in the steam turbine, the steam enters the engine at one end of the working cylinder and exhausts at the other end, whilst the fall in temperature in the working cylinder harmonises, as far as possible, with the conditions present in the steam turbine. It is however, more advisable to utilise the compression, as described above, to raise the temperature at the inlet end in the una-flow cylinder above the temperature at the inlet end in a turbine, and further to lower the temperature of the cold exhaust end in the una-flow cylinder below that of the cold exhaust end in the turbine. This distribution of the temperature, which is emphasised by the effective cover jacketting at the inlet end, produces a flow of heat from the hot inlet end to the cold exhaust end so that, at each point in the cylinder, a temperature is reached which is higher than the steam temperature at the corresponding point in the upper or expansion line of the steam diagram (fig. 4). This distribution of the temperature results in the surfaces of the working cylinder being maintained at a higher temperature than that of the steam to which they are exposed at all points, with the exception of the comparatively short exhaust belt.

The greater part of the heat radiating from the cover is taken up by the steam which remains behind in the cylinder after the exhaust is completed and is used for compression. The heat taken up by the compression steam is then returned to the cylinder head by the piston thereby causing a great increase in temperature. The cycle of events then proceeds anew, that is to say the heat is always pumped back to the cover end by the action of the piston. That the heat conduction from the hot end of the cylinder over the neutral zone to the cold region is very small will be realised from the fact that a blacksmith can hold one end of an iron bar in his hand when the other end is at a white heat. In such a case the temperature difference between the two ends is much greater and yet the amount of heat conducted through the rod to the hand of the blacksmith is very small.

It is advisable to use a steam jacket around the inlet end of the una-flow cylinder when using saturated steam so as to avoid the danger of the actual formation of water drops on the cylinder walls near the exhaust. If the cylinder wall temperature is higher than the temperature of the steam to which the wall is exposed, then water deposits are impossible. The adiabatic action of the cylinder walls, observed and determined by Mr. G. Duchesne, is present when proper jacketting is employed. Mr. Duchesne found that the rate of heat transference between saturated steam and a moist metal surface is 300 times as great as the rate of heat transference with a dry surface at a higher temperature than the steam. This interesting fact explains the remarkable thermal results obtainable from the una-flow steam engine having jacketting which, as shown in practice, maintains the walls at a higher temperature than the steam to which they are exposed, this being effected by a cover jacket only in the case of superheated steam, and by the cover and graduated cylinder jackets in the case of saturated steam. The only reasonable conclusion can be that there is a very limited heat exchange between the steam and the dry hot clearance surfaces in the una-flow cylinder when properly jacketted.

This conclusion is further confirmed by the interesting investigations made by Prof. Hubert and Mr. A. Duchesne on steam engines in the laboratory of the University of Lüttich. These experiments were in the nature of a close investigation into the effects of steam jacketting. Tests were made, first with an engine without jacketting, second with an ordinary steam jacket, and third with a steam jacket supplied with saturated steam at a higher pressure than the working steam. The observed steam temperatures (curve *a*), wall temperatures (curve *B*) and the steam temperature calculated from the pressure-volume diagram (curve *c*) are given in figures 33, 34 and 35, for the three cases mentioned. These results reveal the striking fact that even in the case of a steam engine without a jacket, the steam passing from the cylinder towards the end of exhaust is actually superheated in consequence of the heat transferred to it from the walls, so that the actual temperature of the superheated exhaust steam rises far above the saturated temperature calculated from the diagram and even above the temperature of the cylinder walls. From these tests it is clear that, with a reasonable construction of engine, the walls may be dry at the end of compression. If that is the case in the ordinary counter-flow engine, how much more is it so in the case of the una-flow engine?

Figure 34, shows that superheating commences as soon as the exhaust is opened and that the temperature of the superheated steam rises above the cylinder wall temperature even more than in figure 33.

Figure 35, shows that the cylinder wall temperature is still higher than in the two previous cases and remains practically constant. The steam is superheated during practically the entire cycle. The expansion line is very nearly a true adia-

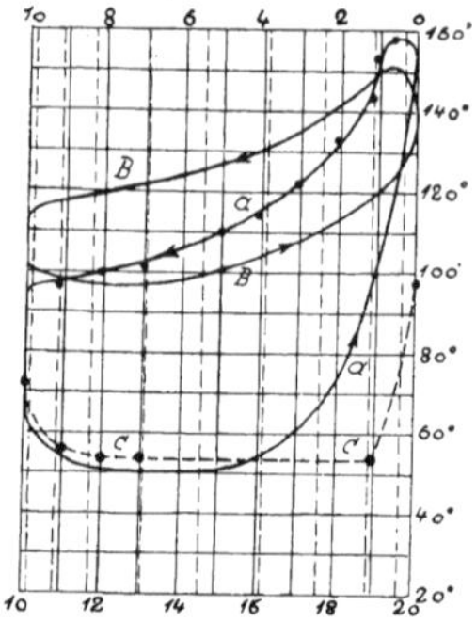

Fig. 33.

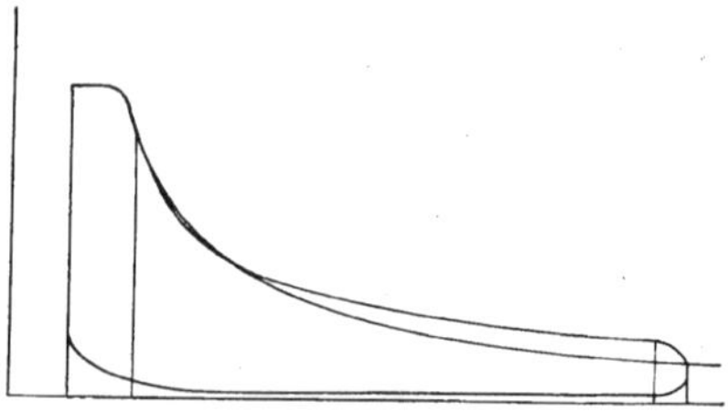

Fig. 36.

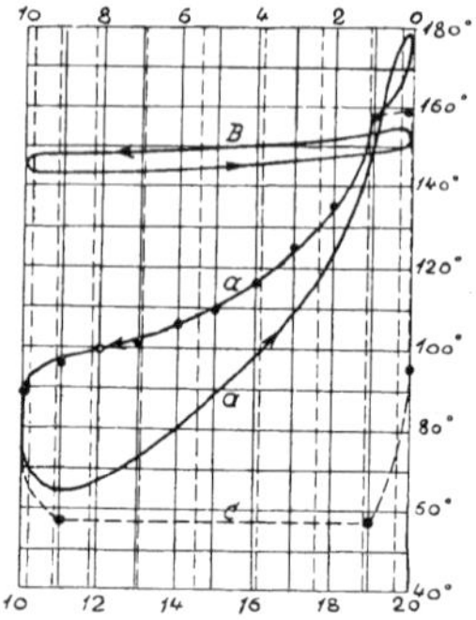

Fig. 34.

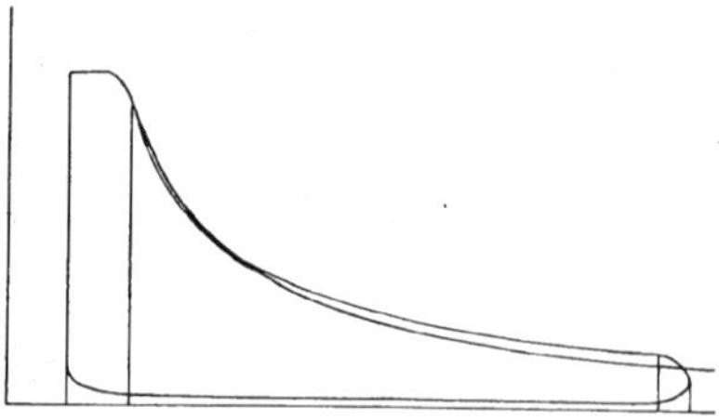

Fig. 37.

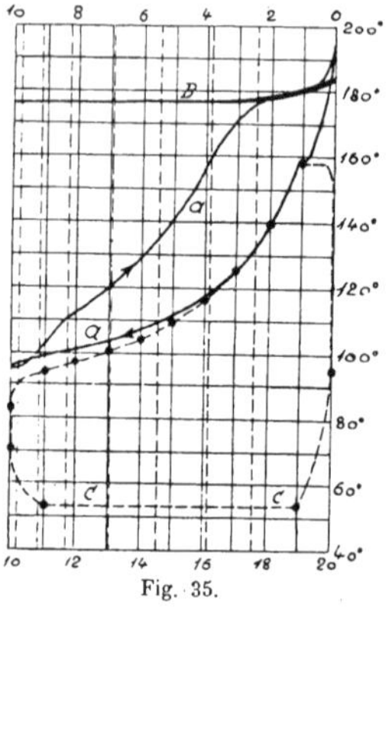

Fig. 35.

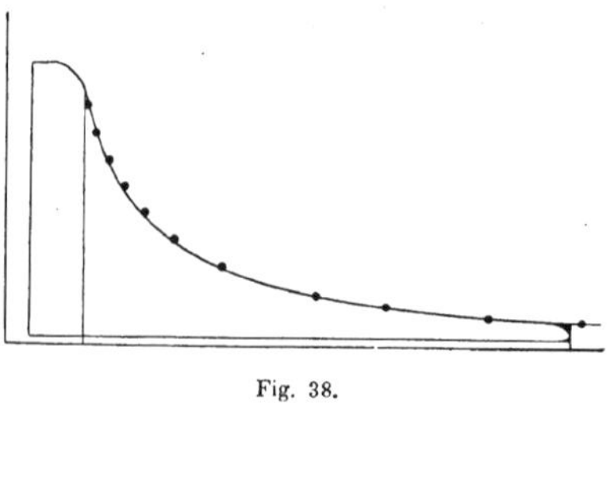

Fig. 38.

batic. The temperature lines during exhaust and compression are very surprising. Initial condensation is entirely absent.

The behaviour of steam as disclosed by these investigations must of necessity produce much more favourable thermal results with a una-flow engine where the fundamental error of the counterflow action is not present. The rapid transference of heat from the large exposed surfaces to the exhaust steam is favoured in the counter-flow engine by the fall in pressure, the high temperature differences, the change in the direction of flow and the high velocity and disturbed nature of the flow of the exhaust taken in conjunction with its protracted duration. The greater part of this superheat in the exhaust steam is pure loss. Although the cylinder condensation is almost eliminated by the intense heating action of the jacket, the amount of heat consumed in the jacket and carried off to be lost in the exhaust also increases. Experience shows that the residue left after subtracting the loss from the gain is almost insignificant. An unfailing test for determining the thermal results obtained in a heat engine is the amount of heat in the exhaust or spent working medium. It has been found that in many steam turbines, the exhaust steam passing from the turbine was superheated — which proves that a portion of the energy of the working fluid has been used up in generating a considerable amount of heat by friction. A similar thermal result is shown in figures 34 and 35, where a considerable amount of the heat in the live steam passes off to the exhaust in the form of superheat and is lost. In order to permit of comparisons, the pressure-volume diagrams for the three tests at the Lüttich laboratory are appended in figures 36, 37, 38 and placed opposite to their corresponding temperature curves. It is interesting to note how the expansion line in cases 1 and 2 (fig. 36 and 37), departs from the true adiabatic whilst it is very approximately a true adiabatic in the third case (fig. 38).

The economic value of the una-flow steam engine is to be traced in the first instance to the uni-directional flow of the steam, whereby a hot end and a cold end are obtained with a gradual fall in temperature between the ends, the temperatures of the walls being at all points higher than the temperature of the steam to which they are exposed, as calculated from the upper line of the diagram. Briefly stated, the cycle of operations in the una-flow engine is as follows:

(a) Admission — no heat flow from jacket;

(b) Expansion — heat transference from jacket to layers of steam nearest to the inlet end, whilst wettest steam follows the piston;

(c) Release — wet steam passed off from cylinder with a scavenging action;

(d) Compression — dry heated steam trapped in the cylinder and compressed up to the inlet end where the clearance is small — heat being transmitted to the steam, during the first part of compression, from the jacket, and the temperature being further increased by the compression.

In this way the reciprocating steam engine is brought to the same thermal plane as the steam turbine, without however possessing the drawbacks of the turbine such as the considerable increase in entropy in consequence of the friction in the nozzles and blades, and the ventilation losses. By adopting the uni-directional flow principle, there naturally follows in its train the omission of the exhaust valve, the absence of large port clearance spaces and surfaces and the leakage losses of the exhaust valve. The working conditions at the centre of the cylinder in the neighbourhood of the exhaust are, from a heat point of view, exceedingly favourable. With an end pressure of 0·7 atmos. and a condenser pressure of $^1/_{10}$ atmos. (absolute in both cases) the cylinder walls immediately adjacent to the commencement of the exhaust ports are subjected at most to a temperature difference of $89{\cdot}5^0$—$45{\cdot}6^0$ C, that is approximately 44^0 C. This temperature difference becomes rapidly smaller towards the centre of the exhaust ports. It increases slowly from the inner edge of the exhaust to the cylinder head. It is also to be noted that the period of time, during which the interior of the cylinder in the neighbourhood of the exhaust ports is connected to the condenser, is extremely small.

The slotted walls proper, through which the exhaust passes, do not really constitute a part of the working cylinder and can have no serious influence on the interior of the working cylinder proper. The maximum theoretical temperature difference between the working medium and the walls occurs at the cylinder head and amounts, in the case of an initial temperature of 300^0 C, and an end pressure of $^1/_{10}$ th atmos. to $300 - 45{\cdot}6 = 254{\cdot}4^0$ C.

This great temperature difference can have no injurious influence, however, because at this point the jacketting and the compression have their most effective action whilst none of the jacket heat is lost. Then again it is at the inlet end that the heat insulating action of the walls, owing to their high temperature, is especially marked, and moreover, the steam particles at the inlet end are practically at rest. In the case of a una-flow engine, those losses, incurred by the high velocity of the exhaust steam over the clearance surfaces in a counter flow engine, are practically avoided. The steam at the cover end is almost always superheated, and being so, is a bad conductor of heat. Another feature of great advantage is that the inlet and exhaust are separated by a considerable distance and that the clearance surfaces are materially reduced.

The clearance space has a deleterious action, first of all because of its capacity or volume and secondly because of the additional surface introduced. The first is small especially if the clearance space is small, but the second point mentioned is of the utmost importance. There is rather more reason to talk of the loss due to the clearance surfaces than to talk of the loss due to the clearance spaces, and much more attention should be paid in designing, to reduce the area of the clearance surfaces. For instance, there is a very serious difference between a 5% clearance space with say one square metre of clearance surface and the same clearance space with two square metres of surface. *The losses due to the clearance surfaces depend on their area, on their arrangement, on the quantity of steam flowing over the clearance surfaces, on the nature of the steam (superheated or saturated), on the temperature differences between the steam and the surfaces, and the amount, duration and nature of the steam current.* The clearance surfaces are considerably increased if they are left in the rough un-

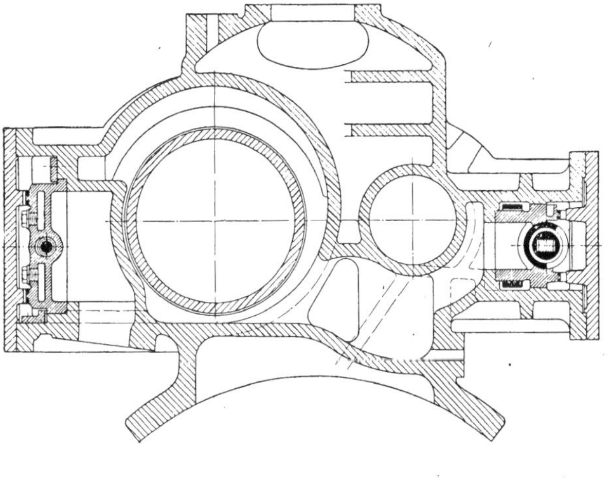

Fig. 39.

machined state in which they leave the foundry. Rough clearance surfaces also affect the nature of the current and the degree of heat exchanged. Clearance surfaces, over which both the inlet and exhaust steam pass, are a source of great

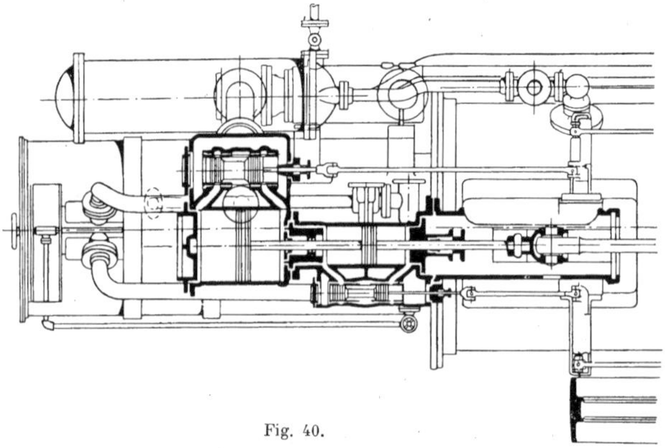

Fig. 40.

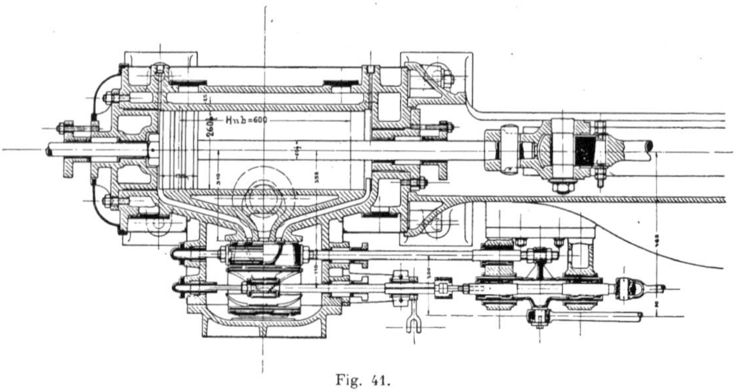

Fig. 41.

loss especially when the temperature differences are great and the exhaust steam is wet. The heat exchanged is considerably reduced when the cycle of operations is carried out wholly under superheat or when the surfaces are maintained by jacketting at a higher temperature than the steam.

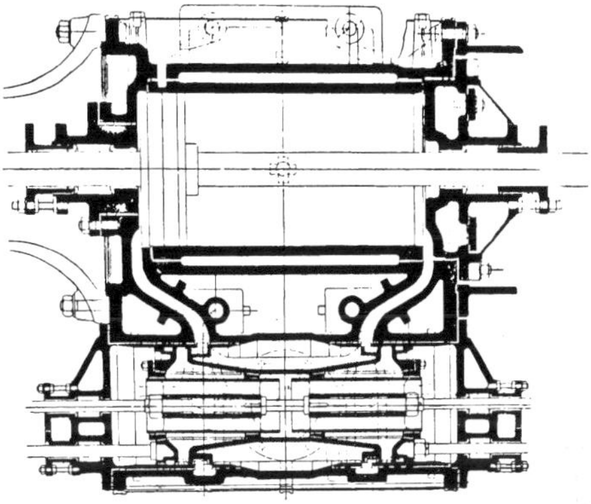

Fig. 42.

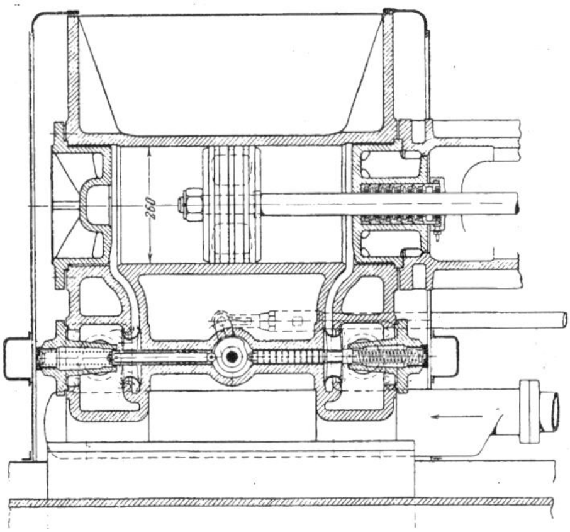

Fig. 43.

The construction of a slide valve cylinder (either for a flat valve or piston valve) as shown in figures 39 to 42, must be especially disadvantageous as in this case the inlet and exhaust steam pass through the same ports.

In this construction, the clearance surfaces (which in most cases are left rough and unmachined), the temperature difference and the duration and amount of the steam

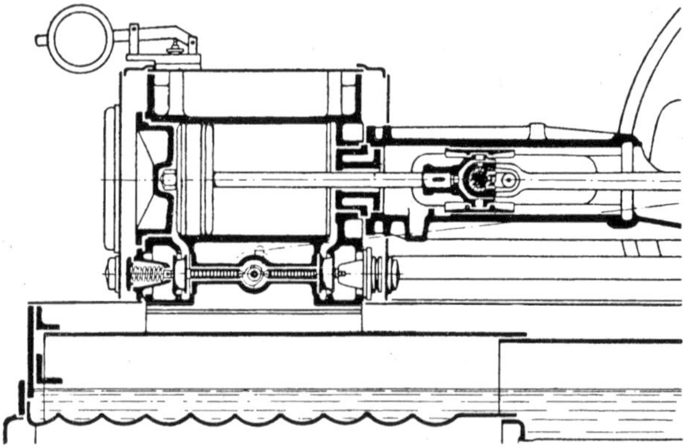

Fig. 44.

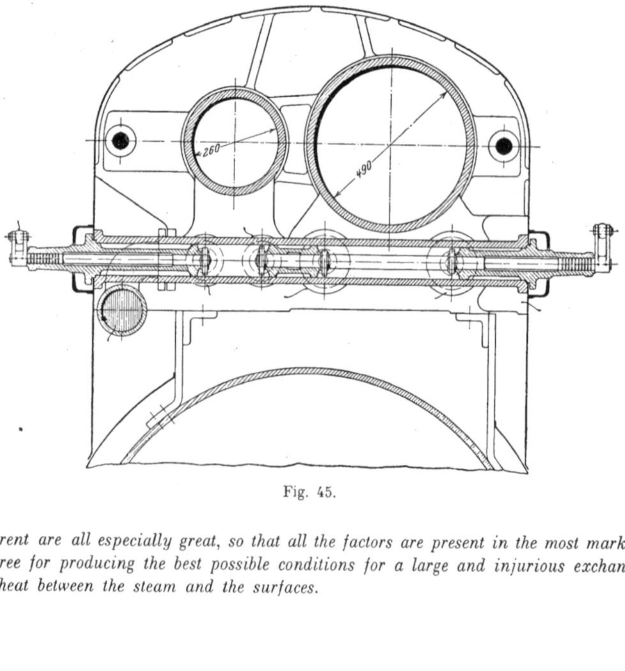

Fig. 45.

current are all especially great, so that all the factors are present in the most marked degree for producing the best possible conditions for a large and injurious exchange of heat between the steam and the surfaces.

The high and low pressure cylinders of a compound engine as shown in figures 43—44 all possess similar faults. In this construction, also, common inlet and exhaust ports are employed. The use of separate inlet and exhaust valves in this case does not materially alter the conditions. The added clearance surfaces are particularly large in the case of the H.P.-cylinder.

The serious defect in all these constructions is confirmed by the importance of superheating as a means for correcting drawbacks therein. There is plenty of scope here for superheating, if it is possible, to correct the serious errors of a defective design.

A good design of steam engine cylinder will prove its superiority most conclusively if it does not require superheating in order to obtain a high degree of efficiency. This is the case in the una-flow engine.

A much better construction than those referred to above, is that shown in figure 46 with a double beat valve, and in figure 47 with a four seat valve, where

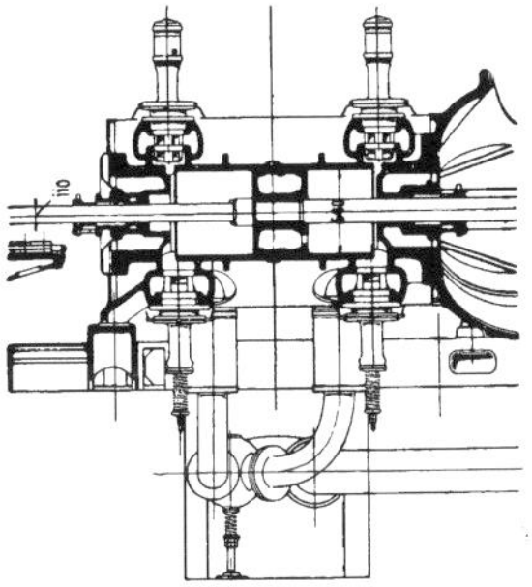

Fig. 46.

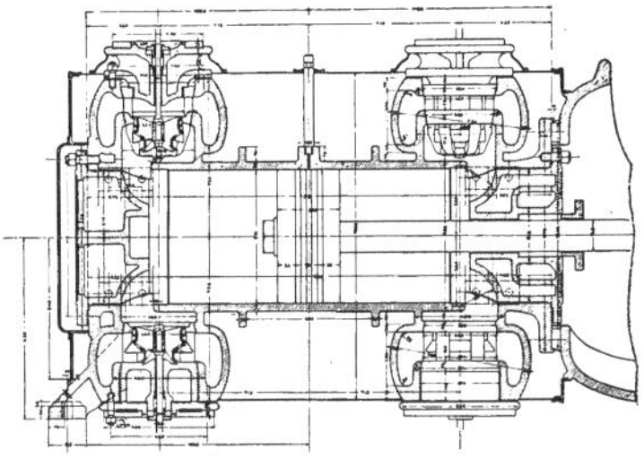

Fig. 47.

the inlet and exhaust passages are divided. In this case the heat exchanged between the steam and the walls is smaller, a fact which is proved by the lower rate of steam consumption obtained by such engines with saturated steam. By omitting the valve casing and arranging the valves in the cover, a considerable improvement could be obtained. The valve casing with the fourseats, in the construction shown, increases the clearance surfaces to a very considerable extent.

The same remarks, with the exception of those directed to the valve casing, apply to the Corliss cylinder shown in figure 48.

The horizontal una-flow cylinder, shown in figure 7, is much better in this respect than the constructions discussed and a still greater improvement is found in the vertical single acting una-flow engine with flat valves which will be more fully described later with reference to figure 102.

In the following table the relative percentages of the added clearance surfaces to double the cross sectional area of the cylinder (that is the smallest possible clearance surfaces at the end position of the piston) are given for various types of steam engine cylinders (figs. 39 to 48, 7 and 102).

Table.

Added clearance surfaces in steam engine cylinders in relation to twice the cross sectional area of the cylinder.

Nature of the cylinder	Percentage
Compound cylinder with flat valves (figure 39)	
H.P.-cylinder, Rider Valve	100
L.P.-cylinder, Trick Valve	80
Tandem cylinder with piston valves (figure 40)	
H.P.-cylinder	200
L.P.-cylinder	120
Ordinary cylinder with Rider valve gear (figure 41)	180
Ordinary cylinder with Rider piston valve gear (figure 42) . .	188
Compound portable engine with lift valve gear (figures 43, 44, 45)	
H.P.-cylinder	185
L.P.-cylinder	103
Ordinary cylinder with double beat lift valves (figure 46) . .	82
Ordinary cylinder with quadruple seated valves (figure 47) . .	110
Ordinary Corliss cylinder, inlet above and exhaust below (figure 48)	
1960 mm diameter, 1500 stroke	70
Ordinary una-flow cylinder horizontal (figure 7)	33
Single acting una-flow cylinder vertical with unbalanced lift valve	
(figure 102)	8·1

When it is considered that the exhaust valve is wholly omitted and the inlet and exhaust are separated by a considerable distance, and the remaining construction is carefully worked out in the case of a una-flow steam engine, it naturally follows that the una-flow engine should stand most favorably in the above table. Another point of immense importance however is that the nature of the clearance surfaces is much superior in the case of the una-flow engine; in fact in this respect the una-flow engine excels all others. This is especially so in the case of the vertical una-flow engine with a single seat lift valve. In designing the una-flow engine, the author has always paid special attention to the arrangement of the clearance surfaces so that they may be easily machined, and he has always prescribed such machining. On comparing the una-flow construction with those referred to above, it will be seen that in most cases machining of the clearance surfaces is absolutely impossible and therefore the total area would come out much greater than the figures given in the table on account of the rough character of the clearance surfaces.

The above table speaks for itself and its message will be specially clear to the thinking reader when he takes into consideration the relative qualities of the clearance surfaces in the various cases. A consideration of these figures should give serious matter for thought in the design of steam cylinders. In the above table, no account has been taken whatever of the increase of the clearance surfaces due to cast on bosses or lugs for indicator cocks, release valves, lubricating connections and similar accessories.

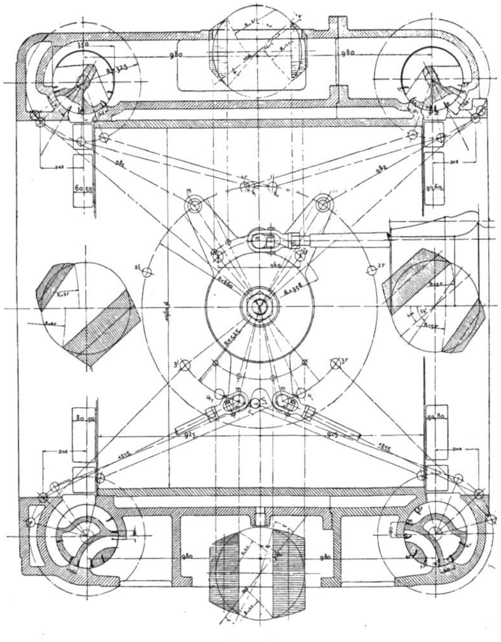

Fig. 48.

The central part of the piston head is protected from the cooling action of the exhaust steam, in the first place by the thin layer of steam which remains immediately in contact with the piston surface. The presence of such a thin layer and its curious action have been verified and determined in an incontrovertable manner by the experiments made in connection with aeronautics. The closing surface, corresponding to the valve face of the ordinary steam engine cylinder, is arranged, in the case of the una-flow steam engine cylinder, on the surface of the cylinder itself. In the case of the counter-flow engine, there must of necessity be a passage connecting the cylinder surface with the valve face, which passage inevitably introduces awkward clearance surfaces. This passage and the corresponding clearance surfaces are wholly absent in the case of the una-flow steam engine cylinder.

Relative velocity between the steam and the outer ring of the piston head only occurs during the first time of the exhaust, when the pressure energy of the steam in the cylinder is converted into velocity energy. This steam current however, makes itself felt mostly in the exhaust ports; that is after it has left the surface of the piston head. In this way the high velocity of the steam during exhaust can have no serious cooling action on the pistonhead. It is only the spent steam

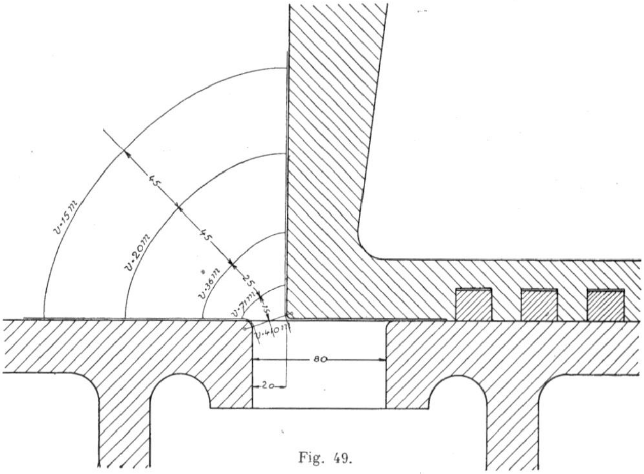

Fig. 49.

which, moving with low velocity towards the exhaust, exercises any cooling action on the piston.

The exhaust ports may be considered as nozzles in which the edge of the piston acts as one of the boundary walls. Treating the exhaust ports as nozzles in this way, and considering the cross section immediately in front of the ports, it will be found that the velocity here is small. The conditions which exist in the case of a cylinder 600 mm Dia. and 800 mm stroke with a port opening of 20 mm, are given in figure 49. On calculating out the velocities, on the basis of normal conditions, it is found that at the narrowest part, the velocity is 410 m per second. At a distance of 15 mm from the narrowest part, the velocity is 71 m, at 40 mm it is 36 m, at 85 mm it is 20 m, and at a distance of 130 mm away from the narrowest point the velocity is only 15 m.

It should be noted that a reduction in the cross section occurs at the narrowest point owing to the metal webs between the individual exhaust ports, and these webs form part of the surface over which the steam passes on its way to the exhaust. As the piston advances over the exhaust ports and uncovers more of the web area, the relation of the various cross sections is such that the velocity of flow of the steam towards the exhaust ports over the cylinder surfaces is increased, but, it is also to be taken into account that during the time the piston is moving over the exhaust ports, the pressure fall in the "nozzles" is less so that the velocity of flow of the steam in the cylinder towards the exhaust ports is really decreased.

According to the diagrams, the exhaust appears to be completed when the crank is at its dead point, so that the amount of the steam current and the duration of its flow over the piston surface are very small as also will be the cooling action resulting from such flow of steam. The entire cross section of the cylinder acts as the preliminary conduit of large cross section leading up to the nozzles formed by the exhaust ports. The quiescent layer of steam adhering to the piston, the energetic heating of the piston head during compression and admission all co-operate to prevent excessive cooling of the piston. Finally, it is only necessary to refer to tests of the una-flow steam engine to show indisputably, that this engine stands on the same high level with the best compound and triple expansion steam engines. This can only be possible, if the cooling action of the piston is of little practical importance.

One of the best features of the una-flow engine is its high compression. The transference of heat from the cylinder or cover jacketting is promoted considerably by the great temperature differences which arise. The compression, starting immediately after the most rapid heat transference, brings this heat to act in a manifold intensified degree on the small clearance surfaces. Upon this action there depends quite a number of important features. It is due to this that the temperature at the end is so high and also that the internal temperature of the cylinder walls is in excess of the temperature of the working steam at every point. This temperature condition of the walls ensures the adiabatic action. Naturally the designer of the engine may modulate the compression as he desires by suitably increasing the clearances.

High compression increases the mechanical efficiency of the engine. In this connection reference is made to figures 51, 52, which show how admirably the effective effort is distributed over the entire piston stroke. This is in part the cause of the high mechanical efficiency of the una-flow engine which reaches the remarkable figure of 95% (with a mean effective pressure of 3 atmos. and forced lubrication).

Chapter VI.

Una-flow stationary engine.

The una-flow steam engine for general land power purposes may be used as a condensing engine or with exhaust to the atmosphere, and it may work with either superheated or saturated steam. The form of the diagram (fig. 50, 51, 52) shows that high speeds and high piston velocities (between 4 and 5 metres, per second) are attainable.

If the inertia line and the effective piston pressure line are drawn together, it will be seen that the resulting diagram shows an almost uniform effort during the entire stroke. If the highest ordinate in this diagram is taken and used for calculating the stresses in the driving mechanism, it will be found that these stresses are higher for the una-flow engine than for a tandem engine of equal power.

Figure 53, shows the actual difference between the tandem and una-flow steam engine diagrams. If the same area of diagram are to be obtained, a smaller una-flow cylinder as compared with the L.P.-cylinder of the tandem engine will suffice, on account of the more complete nature of the diagram in the case of the una-flow engine. In the case taken the ratio of the two cylinders is, 1: 1·075. Taking this relationship and the different values given by the comparative diagram shown in figure 53 (that is the diagram where the inertia forces are neglected), it is found that the ratio of the piston rod pressure in the una-flow to that in the tandem engine is 2 · 15 : 1. This ratio falls to 2 : 1, and 1 · 8 : 1, when the cut-off in the high pressure cylinder is later by 50% to 100%. The stresses on the working parts in a tandem engine increase with increase in the load (the cut-off in the L.P.-cylinder remaining constant) whilst the maximum stresses in the una-flow engine remain the same. It must be taken into account, however, that it is not the maximum steam pressure which determines the actual stresses, but the greatest intercept in the diagram shown in figure 52, where the effective steam pressures and the inertia curves are shown in one diagram. In most cases this will give a much better show in favour of the una-flow engine.

A very good compound engine has been introduced into the above comparison. Considering usual constructions, the ratio of the two cylinders will be 1 : 1·1 to 1 : 1·13.

In the diagrams shown in figures 54 to 57, the power fluctuation curves for calculating the flywheel of a una-flow steam engine, and the same curves for the case of a tandem engine of the same power are shown. Figures 54 and 56 show the curves for the una-flow engine, and figures 55 and 57 those for the tandem engine. The una-flow engine has a cylinder of 600 mm diameter and 800 mm stroke, and runs at 155 revs. per minute. The tandem engine has a H.P.-cylinder of 430 mm Dia, and L.P.-cylinder of 725 mm Dia, a stroke of 800 mm, and rotates at 125 revs. per minute. The speed of revolution is so arranged in relation to the reciprocating masses in each engine that the same uniform balancing is obtained in both cases.

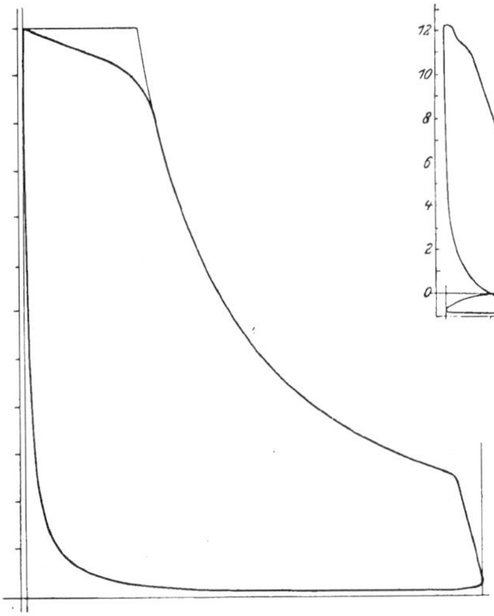

Fig. 51.

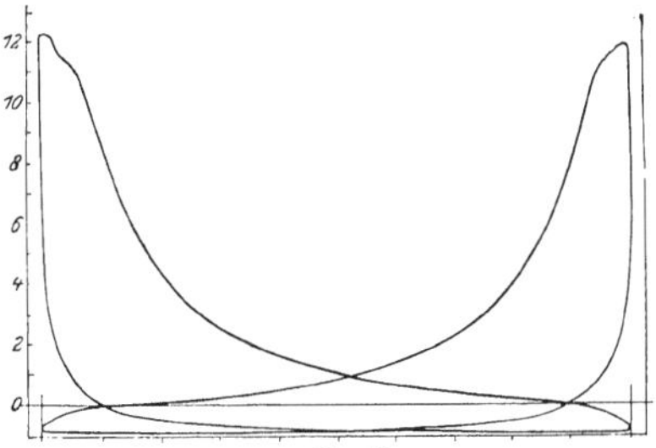

Fig. 50.

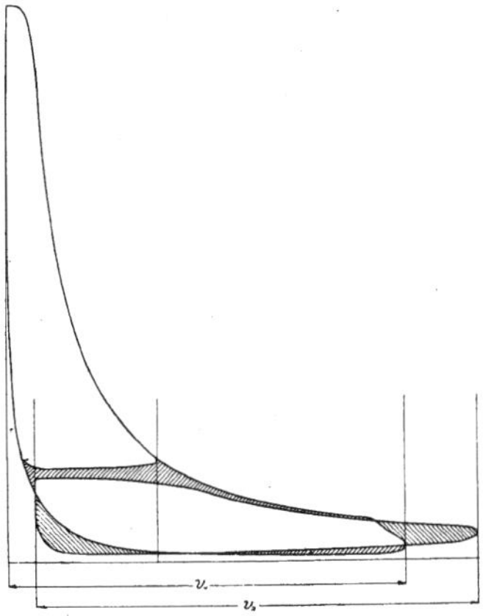

Fig. 53.

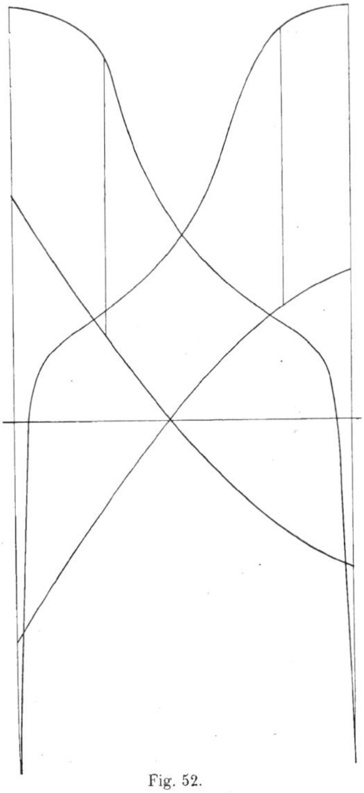

Fig. 52.

The una-flow engine shows $\varphi = 0{\cdot}108$, and the tandem engine $\varphi = 0{\cdot}133$. These figures represent the ratio of the greatest excess area, above the mean tangential effort line to the total area of the tangential effort diagram. The two figures are in the ratio approximately of $1:1{\cdot}33$, that is to say the weights of the two fly wheels are in the ratio of $3:4$.

These figures, however, totally neglect the differences in the speed of revolution. Taking the speed of revolution into consideration, it will be found that there is a

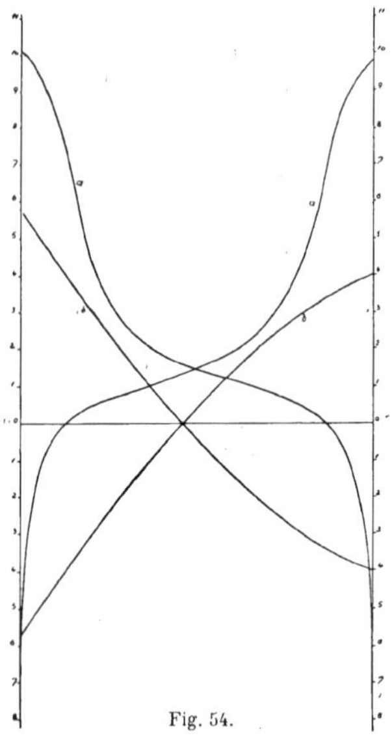

Fig. 54.

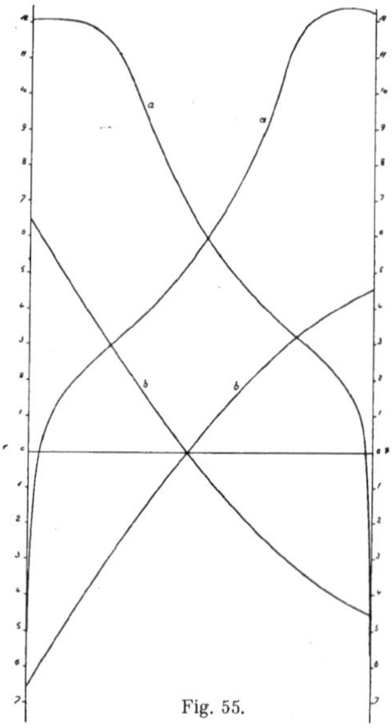

Fig. 55.

still further material reduction of the fly wheel weight in favour of the una-flow engine. The fly wheel in the case of the una-flow engine should be heavier, if it is to be sufficient to enable the engine to work smoothly with exhaust to the atmosphere. This is also the case when the engine is to work steadily at overloads. The losses due to throttling between the cylinders and between the last cylinder and condenser, as well as the additional radiation loss consequent upon the use of two cylinders, an intermediate receiver and connections, are all inevitable sources of loss in multiple expansion steam engines, whereas these sources together with the losses, are all avoided in the una-flow steam engine.

In order to enable the engine to work with atmospheric exhaust, when required, extra separate clearance chambers are arranged in the cover. These chambers are adapted to be put in operation or cut out as required and are preferably located at the side of the cover remote from the cylinder, so that during the time the engine is working with condensation they are cut out of operation and act as heat insulating chambers. When working with exhaust to the atmosphere,

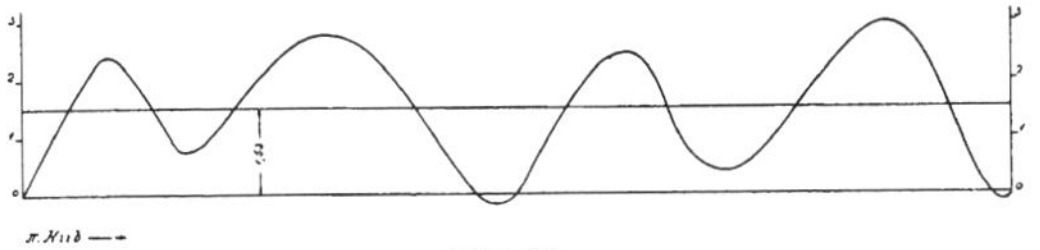

Fig. 56.

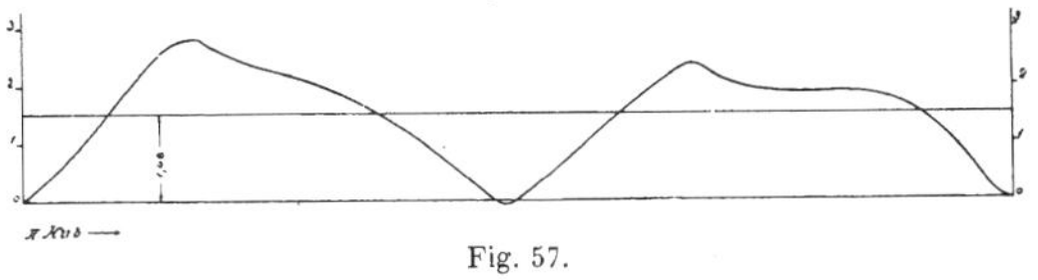

Fig. 57.

the additional clearance chamber is put into operation and has then the effect of doubling the cover heating surface. *Every una-flow stationary condensing steam engine for power purposes, should be provided with an additional clearance chamber which may be cut out of operation when desired, so that starting of the engine may*

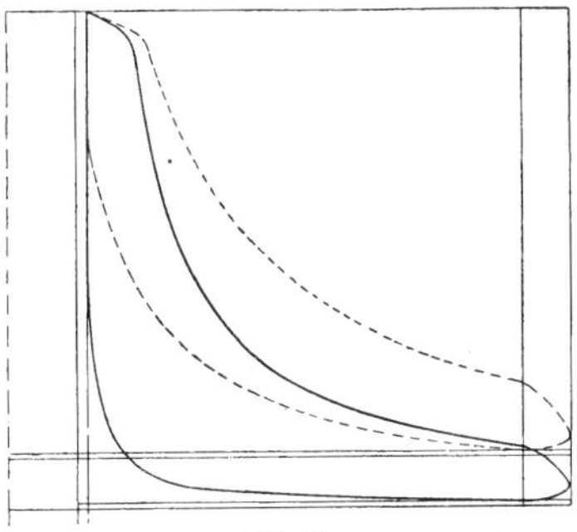

Fig. 58.

be facilitated in cases where the air pump of the condenser is driven from the engine and the engine may work with exhaust to the atmosphere in cases of emergency.

This additional clearance space is a chamber connected to the inlet end by a passage which may be opened or closed by a valve when required. (See fig. 77—80.)

Fig. 58 shows two indicator cards, one taken whilst starting, when the additional clearance space is in operation, dotted lines, the other in ordinary working condition, with the additional clearance space shut off.

The clearance space, when working with condensation, is on an average with a good vacuum from 1½% to 2%, and when the additional clearance space is put into communication with the cylinder the total clearance is from 12% to 16% according to the initial pressure.

When the engine is to work with exhaust to the atmosphere, the piston may be concave at its end and the concavity should be such as to accommodate a total clearance of 12% to 16%.

The range of cut-off in the "una-flow" engine may be from nil to one quarter of the stroke. The governor and the valve gear have to be constructed for this range. The governor and gear must also operate to give the inlet valves absolutely no opening in the extreme position. It must also be

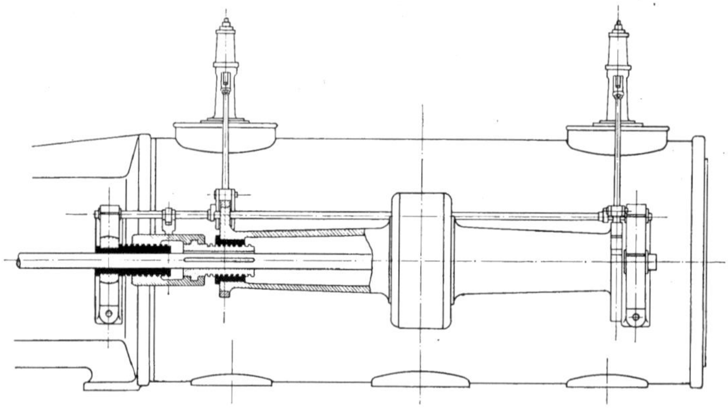

Fig. 59.

borne in mind that the admission should occur, as near as possible, at the commencement of the working stroke. If the admission is given any material advance, it has been found that with a condensing una-flow steam engine, the running is not so sweet and easy. In the case of una-flow engines with exhaust to the atmosphere the running is always smooth.

One form of una-flow engine valve gear is shown in figures 59—60, in which a special valve operating shaft is provided and a shaft governor is arranged to control the shifting eccentric by a small relative rotation about the centre of the shaft to change the angle of advance. This gear does not operate in the ordinary manner by adjusting both the stroke and angle of advance. The stroke remains constant and the outside lap is altered by an amount proportional to the alteration in the angle of advance. A centrally pivoted swing lever is interposed in the connecting gear between the eccentric and the valve, and the central pivot of this lever is mounted on a cranked part of a rod passing over the width of the cylinder so as to act simultaneously on both valves. This rod may be carried in brackets cast on the brackets which carry the valve operating shaft. When the main eccentric is set to a different angle of advance by the governor, the adjustment of the eccentric causes it to screw on the threaded sleeve shown in section on the left hand of figure 59. This threaded sleeve is so mounted as to slide on a feather on the valve-operating shaft, and it is coupled to the lap-adjusting sleeve which engages a quick-pitched thread (shown in black) fixed rigidly in one of the bearings of the valve-operating shaft. When the lap-adjusting sleeve is moved axially over the fixed quick-pitch screw, an arm carried by the sleeve rotates and this rotation is transmitted by the short connecting rod and crank, seen in figure 60, to the

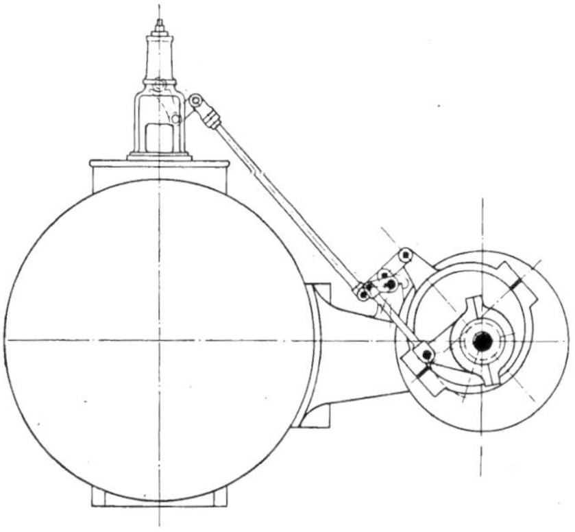

Fig. 60.

cranked rod which carries the swinging levers interposed in the valve operating gear. The pivots of the swinging levers are thus moved in the arc of a circle, the amount of the motion and the radius of the path being related to and dependant upon the amount of the variation in the angle of advance and the eccentricity of the eccentric. When the swinging lever is centrally pivoted, the radius of the arc of motion of the pivot is a half of the eccentricity of the eccentric. When the main eccentric is at the dead point, the pivot of the swing lever is at the same point. This is also the case with angles of 30°, 45° etc. In this way the required alteration in the lap for each different position of the shifting eccentric is obtained. This removes one of the serious defects of the usual valve gears in which the designer is compelled to take a small eccentric throw and small valve lift only for being able to give occasionally a large cut-off. Hereby great throttling losses are involved at small cut-offs. The valve movements are always the same with a valve gear such as described. The cut-off may also be easily made as late as 90% of the stroke, whilst the point at which admission commences may be approximately the same for all loads. A reversing gear acting on the same principle is described later in the chapter on the Marine una-flow engines.

The arrangement of the entire valve gear above the bed plate will be found most convenient whilst it leaves the entire space in the foundations free for the pipes and the close connection of the condenser to the exhaust belt (figs. 7 and 8).

The Erste Brünner-Maschinenfabriks-Gesellschaft decided first of all to alter an old 80 H.P. single cylinder condensing engine with a forked frame (Fig. 61). The

Fig. 61.

old counter-flow cylinder was replaced by a una-flow engine cylinder 400 mm dia. and 420 mm stroke designed by the author.

A shaft governor and shifting eccentric were fitted on the free end of the main shaft and the valves were driven from this eccentric through a rock shaft mounted on brackets on the exhaust belt. Two Lentz cam operating gears were employed for the inlet valves. Even although this patched up construction was defective and the condenser vacuum not too good, the same steam consumption per HP was obtained as given by compound engines of the same size. Figure 62 shows another una-flow engine built by the same makers.

A 300 HP una-flow steam engine built by the Gebrüder Sulzer of Winterthur and now in operation in the copper works of Gebr. Wieland in Ulm is shown in figures 24, 63 and 64. The shaft governor for this engine is shown in figure 65, whilst a cover with inlet valve and valve bonnet are shown in fig. 66.

Fig. 62.

Fig. 63.

Fig. 64.

Two una-flow engines of 550 H.P., each as shown in figure 67 were built by Gebr. Sulzer and are working at present in the Haford copper works in Swansea England. A longitudinal and a cross section of this engine are shown in fig. 68.

Another una-flow engine of 350 H.P. built by Sulzer for the cotton-mill Lampertsmühle near Kaiserslautern is shown in figure 69. Another una-flow engine of about the same size was built by Sulzer for the Gußwaren-Actiengesellschaft in Kaiserslautern.

All these engines were highly successful, the steam consumption in each case being in full accordance with the results reported about above. This

Fig. 65.

Fig. 66.

was also substantiated by Mr. Steiner, Chief mechanical engineer of Gebr. Sulzer, on the occasion of a meeting of the society of German engineers at Stuttgart when he said: *We have found that, as regards steam economy the efficiency of the una-flow engine is not merely equal to that of the compound engine but is actually higher.*

A una-flow steam engine built by the Görlitzer Maschinenbau-Anstalt is shown in figure 70.

Shortly after the attempt of the Erste Brünner Maschinenfabriks-Gesellschaft, the Elsässische Maschinenfabrik decided to make an experiment on a larger scale. They built a 500 H.P. engine (640 mm dia. and 1000 mm stroke) for electric generating

Fig. 67.

Fig. 68.

Fig. 69.

purposes (figs. 23 and 71). This engine proved a decided success. It was subjected to a test lasting for four hours and eight minutes by the Alsace Society of Steam boiler owners (Elsässischer Verein der Dampfkesselbesitzer). The results were as follows:

1. Steam engine.

Mean pressure in front of starting valve	12·60 kg per cm²
Mean admission pressure of the cylinder	11·90 kg per cm²
Mean steam temperature in front of the starting valve	331·0° C
Mean steam temperature in front of the inlet valve .	305·0° C
Exhaust pressure in the cylinder	0·145 kg per cm² abs.
Steam pressure in front of oil separator	0·121 atmos. absolute
Steam pressure in the condenser	0·075 atmos. absolute
Revolutions per minute	121
I. H. P.	503·1
B. H. P.	465·7
Mechanical efficiency	92·5°
Steam Consumpt. per H.P. hour	4·6 kg
Temperature of the spraying water in the condenser.	12·0° C
Temperature of the water after leaving the condenser	31·1° C
Weight of condensing water for 1 kg of steam . . .	30·0 kgs
Amperes to driving motor for condenser	50·7 amps.
Volts to driving motor for condenser	244·5 volts
Power to the driving motor for the condenser	12·4 KW

2. Dynamos.

Amperes	1277·0 amps.
Volts at the brushes	250·1 volts
KW at the brushes	319·4
Efficiency of the dynamo (electrical)	93%
KW at switchboard	307·0 KW
Steam consumption per useful KW hour	7·55 kg

In the above tests the pipe connections to the condenser were too small and there was an oil separator placed between the exhaust outlet and the condenser, both of which factors had a serious effect on the back pressure, which was 0·145 atmos. If the condenser had been arranged as described above, the cylinder jacketted, and proper steam tight valves used, the above results would have been materially improved. As a matter of fact the steam consumption for a modern una-flow engine under favourable circumstances is 4 to 4·1 kg per 1 H.P. hour.

A 900 H.P. una-flow engine built by the Elsässische Maschinenfabrik-Gesellschaft of Mülhausen has already been shown in figures 25 and 26. Attention is specially drawn to the strong fork shaped engine bed, to which several firms give the preference, whilst others still adopt the overhung crank. A good deal may be said in favour of the double crank on account of the high piston pressures especially in the case of short stroke engines.

Figure 72 shows another engine built by this firm. Figure 73 shows the governor and parts thereof. Both figures exhibit clearly the great simplicity of the design.

A mill engine built by Burmeister & Wain of Copenhagen, is shown in figures 74 and 75. Special attention is drawn to the immediate connection of

Fig. 70.

Fig. 71.

the condenser to the engine and to the method in which the end of the steam engine cylinder is supported on two adjustable struts. As the engine is intended to work with high superheat, the cylinder is not jacketted (see figure 76). The cover jacket is extended over that part of the cylinder uncovered by the piston during normal cut-off. The piston is reduced in diameter so as to act like a plunger for a short distance at its end. This reduction of the end part of the piston allows for the greater expansion which occurs at this point. The additional clearance spaces for starting are connected to the cylinder by passages on each side of the inlet and are controlled by valves (see figure 77—80). One valve is sufficient for starting the engine when the air pump of the condenser is driven by the engine. The second valve is provided to enable the engine to work at full

Fig. 72.

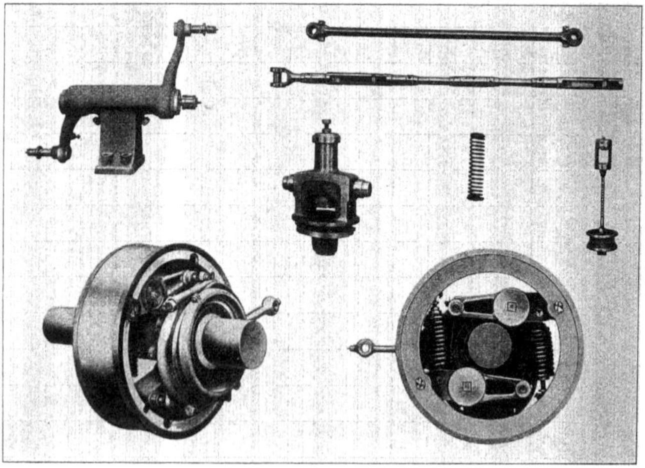

Fig. 73.

speed without condensation. The cover has the minimum amount of contacting surface with the engine bed, in order to avoid, as far as possible, heat conduction losses (see figure 77). The casting is entirely symmetrical so as to diminish the cost of patterns. The results of a test made on Feb. 5, 1910 by Mr. H. Bache, Prof. Tech. High School at Copenhagen, with this engine are given below and should prove most interesting. The report on this test was dated Feb. 16, 1910. In this work Prof. H. Bache was assisted by Mr. O. Flamand.

Fig. 74.

Power			Pressure in kgs/cm²	Steam Temp. C	Vacuum in °/₀ of 760 mm.	Revs. per min.	Steam consumpt. per hour kgs.	Steam consumpt. per hour per		
K.W.	B.H.P.	I.H.P.						K.W.	B.H.P.	I.H.P.
64·50	98·7	116·0	9·90	352	94·0	179·0	479·0	7·44	4·86	4·12
86·46	130·0	149·0	9·87	354	93·8	175·0	632·0	7·32	4·86	4·24
108·66	163·0	184·5	9·84	353	93·5	176·5	798·4	7·36	4·90	4·34
131·24	197·0	222·0	9·80	353	92·6	173·5	976·7	7·45	4·97	4·40
109·00	164·0	186·0	9·75	saturated	93·0	178·0	1150·6	10·55	7·03	6·20

The above results come fairly close to those quoted in Chapter 3, for a 300 HP una-flow steam engine. They are not quite so good as the results given in Chapter 3, but this engine is not so large as the 300 HP engine. The cylinder in the tests given above was 450 mm dia. and 500 mm stroke, whilst the 300 HP engine cylinder was 600 mm dia. and 800 mm stroke. Then again there was no jacket around the cylinder in the smaller engine. This jacket also even when disconnected shows a beneficial effect on the consumption of saturated steam as will be clear by comparing the results of both engines.

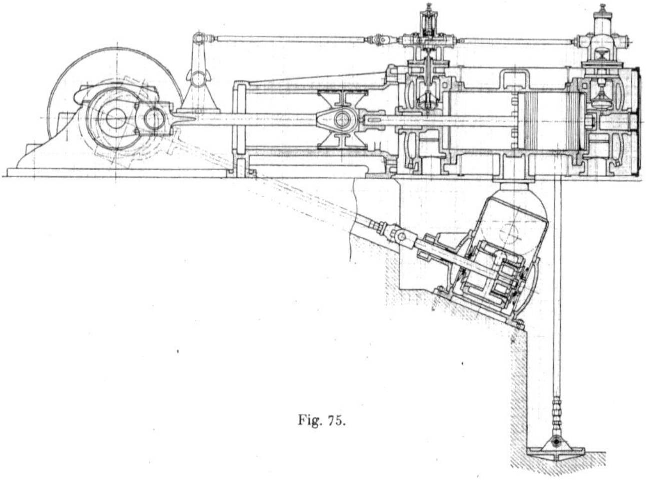

Fig. 75.

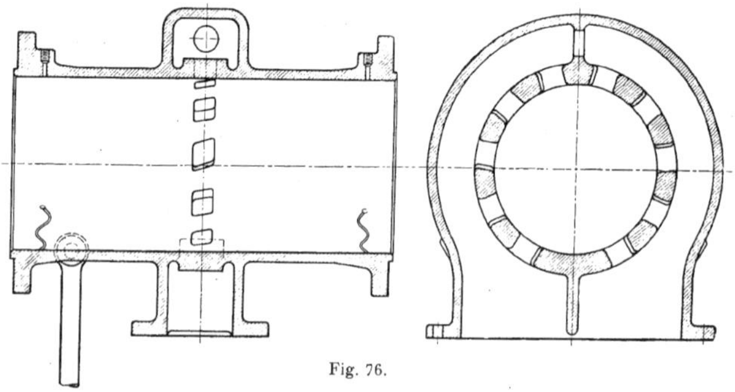

Fig. 76.

Messrs Burmeister & Wain have resolved to adopt the cylinder jacket more generally on account of the material difference in steam consumption which it produces.

Figure 81, shows a type of single acting una-flow steam engine built by Messrs Burmeister & Wain of Copenhagen, which has been adopted by many of the Danish Dairies. The valve drive is arranged horizontally at the side, so that the cam rod for the valve may be operated directly from the shifting eccentric (see figures 82 and 83).

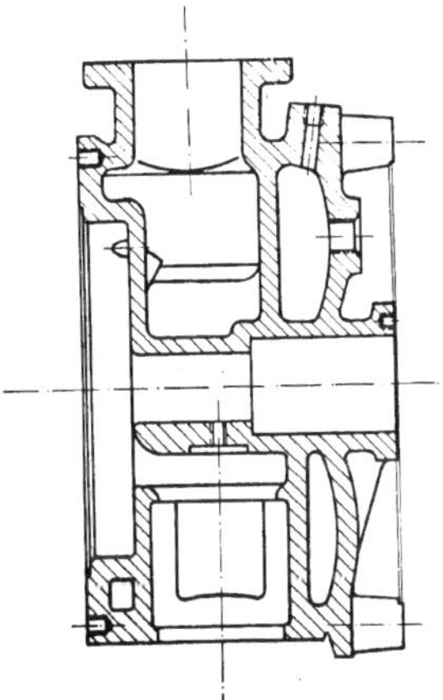

Fig. 77.

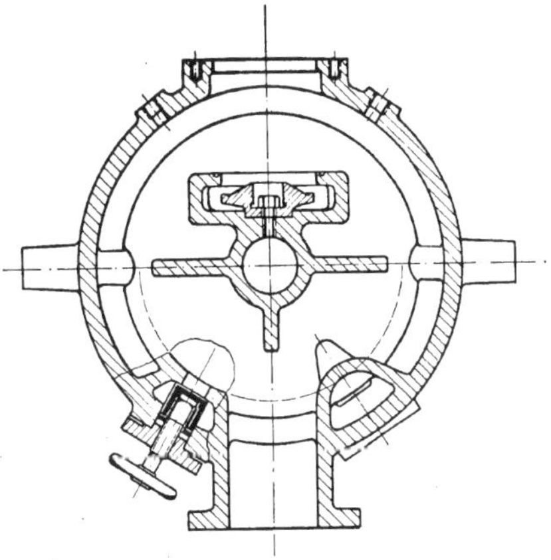

Fig. 78.

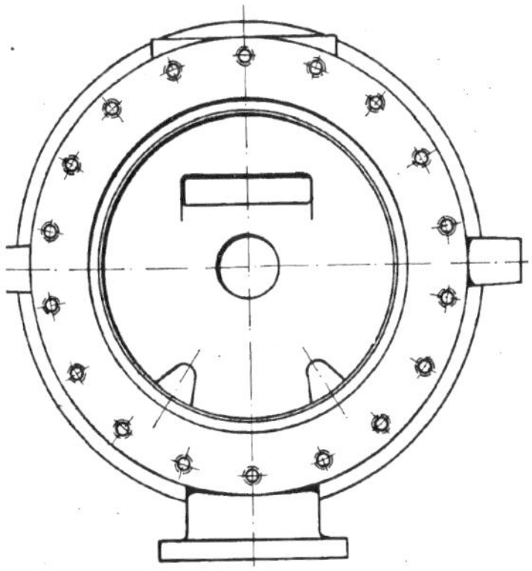

Fig. 79.

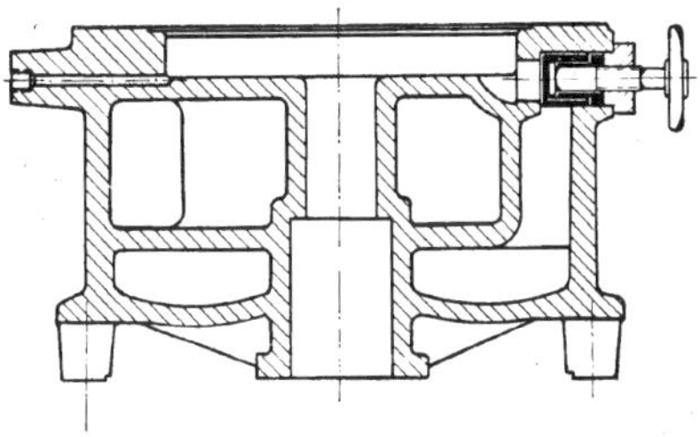

Fig. 80.

A una-flow engine built by Messrs Ehrhardt & Sehmer of Saarbrücken is shown in figure 84. The engine is running at the central station of the Saartal electric railway in Saarbrücken. This engine proved entirely satisfactory. As regards its behaviour under overloads, it far surpassed all expectations. It worked at almost double its normal load for quite a long time. With a normal cut-off at 8% of the stroke, the range of possible overloads in a una-flow engine is larger than can be obtained in other known forms of engines. For, if the rods and mechanisms are calculated for the initial pressure, there is no increased stress placed on the rods due to the overload.

Fig. 81.

The engine illustrated in figure 84, has the following dimensions.

Cylinder diameter	650 mm
Stroke	1000 mm
Revs per minute	130
Normal load	500 HP

This was the *first* engine built by Messrs Ehrhardt & Sehmer and the following gives the results of tests made with it.

Duration of test	6 hours
Steam pressure at the engine	10·1 atmos.
Steam Temp. at the engine	251° C
Temperature of condensing water	25·1° C
Vacuum in condenser	66·6 cm
Barometer	74·7 cm

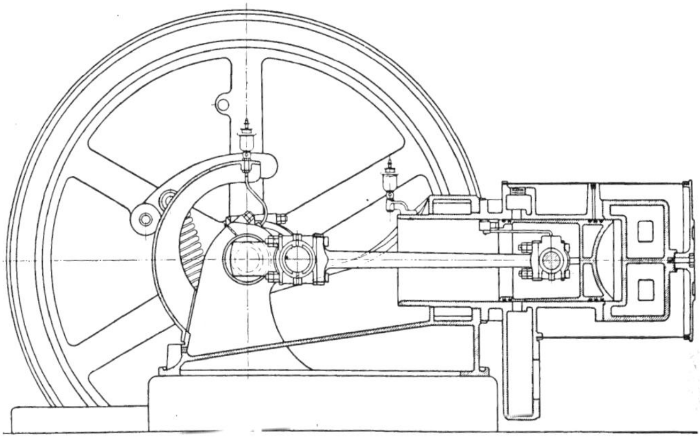

Fig. 82.

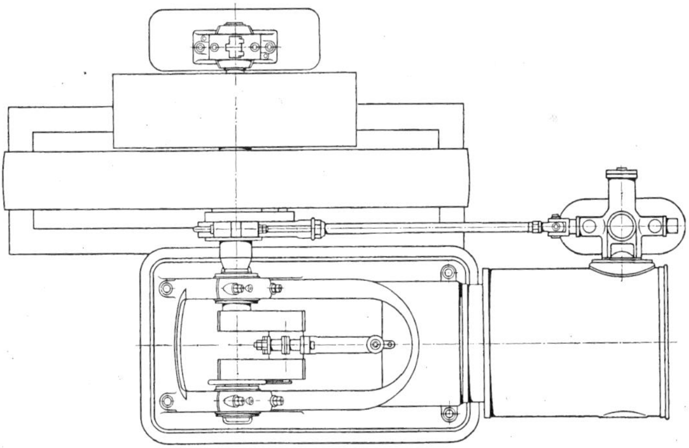

Fig. 83.

Vaccuum in % of Barometer	89%
Revs. per min.	131
I. H. P.	386 H.P.
Output of D. C. dynamo	231 KW
Guaranteed efficiency of D. C. dynamo	92%
B. H. P. of engine	341 H.P.
Mechanical efficiency of the engine	88·2%
Total steam consumpt. per hour	1915 kgs.
Steam consumpt. per I. H.P. hour	4·95 kgs.

In examining the above results, it should be noted that there was no jacket over the cylinder. At the time of building this engine, the data from experiments with cylinder jacketting was not to hand. By using a steam jacket round the cylinder the steam consumption per I.H.P. could be reduced by about ½ kg.

Fig. 84.

A una-flow stationary engine (400 H.P.) built by the Maschinenfabrik, Grevenbroich is illustrated in figure 85. This engine runs at 175 revs. per minute. A 500 H.P. una-flow engine, built by this firm for the Margarine factory of Messrs Jürgens & Prinzen of Goch, is illustrated in figure 86.

Figure 87, shows another 400 H.P. una-flow engine built by the Maschinenfabrik Grevenbroich for the Gewerkschaft Bonifazius of Westig (Westf.). As can be seen from the picture, all the engines built by this firm have the roller and cam valve gear, the sliding rod being operated from the main shaft. There is no tail rod on the piston. The original construction is still adopted and the experience of this firm is that this construction answers all requirements in smaller sizes. For larger sizes the layshaft for operating the valves is preferred.

Figure 88 shows an engine built by Messrs. Musgrave & Sons Ltd., of Bolton, England. This firm has had signal success with the una-flow engine in England.

The steam consumption with a 500 H.P. una-flow steam engine built by Messrs. Musgrave & Sons Ltd., was 4·5 kg per I. H. P. hour. The steam pressure was 11 atmos., the temperature 240° C and the load was a constantly varying one, as the engine was set to drive a tin rolling mill. Considering these unfavourable conditions, the result mentioned is highly satisfactory. There was no jacket around the cylinder.

Fig. 85.

A una-flow engine built by Messrs. Musgrave and installed at Bradford was very carefully tested and gave the following results:

Cylinder diameter	775 mm
Stroke	990 mm
Revs. per minute	120
Steam pressure at stop valve	8·2 atmos.
Steam temperature at the stop valve	254° C
Mean diagram pressure on the crank side	2·5 kgs. per cm²
Mean diagram pressure on the rear side	2·18 kgs. per cm²
I. H. P.	570
Temperature of the condensing water	24·4°
Vacuum in the condenser	90%
Steam consumpt. per I. H. P. hour	4·93 kgs.

If the cylinder had been jacketted there would have been a considerable reduction in the steam consumption.

Two large una-flow steam engines built by the Vereinigte Maschinenfabrik Augsburg-Nürnberg are shown in figures 89—91. Details are shown in figure 92. This firm uses a lay-shaft, driving from this lay-shaft horizontal valves by means of cams. The valve for switching in the additional clearance spaces is arranged at the opposite side to the inlet valve. This firm has also introduced special arrangements for bringing the extra clearance space into operation automatically when the condenser suddenly becomes inoperative. The same result is obtained by employing partially balanced admission valves or spring loaded switching valves for the additional clearance space, these valves acting as safety valves.

By arranging the valve operating shaft at the side, the operation of the valve gear is more correct, because the valve gear is not subjected to the action of the

Fig. 86.

Fig. 87.

cylinder expansion. Care is to be taken that the admission starts in the dead point, when working with a condeuser.

When employing a reciprocating cam rod driven from the main shaft, the longitudinal expansion of the cylinder has to be taken into account. When, as is preferable, the cylinder walls are jacketted and the jacket is heated by a special connection from the steam supply pipe in front of the stop valve, the engine may always be heated up before starting so that the valve gear may always be set

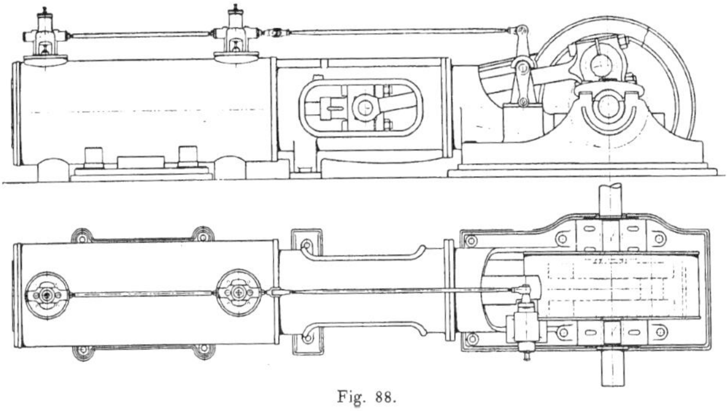

Fig. 88.

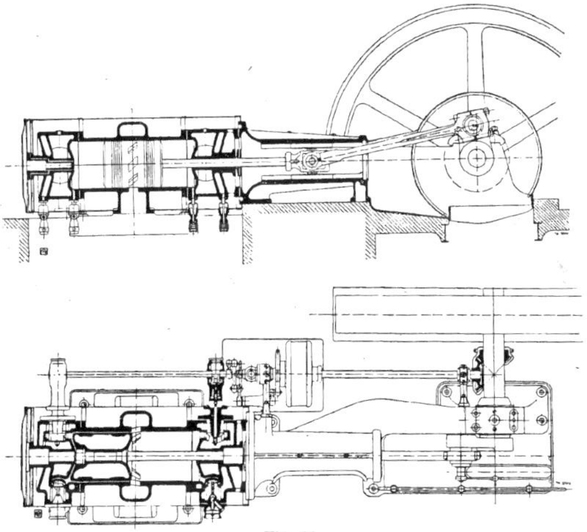

Fig. 89.

exactly to meet the conditions which prevail when the cylinder is heated up. When the engine cylinder is not jacketted, it is only to be expected that, during the first time of running, the valve gear will not act correctly, and it will only commence to act correctly when the cylinder is fully heated and expanded. In all cases it is best to provide the valve gear with an outside lap so that the admission is a little late during starting, when the cylinder is not completely ex-

Fig. 90.

Fig. 91.

panded. It is the lesser evil in the case of the una-flow engine to make the admission too late rather than too early for the reason that, with admission too early, the engine does not run smoothly. The irregularity of the cut-off at starting and some change in speed must however be expected.

When the cylinder is not jacketted, it is still possible to obtain equality in the expansion of the valve rod between the two valve bonnets by enclosing the valve

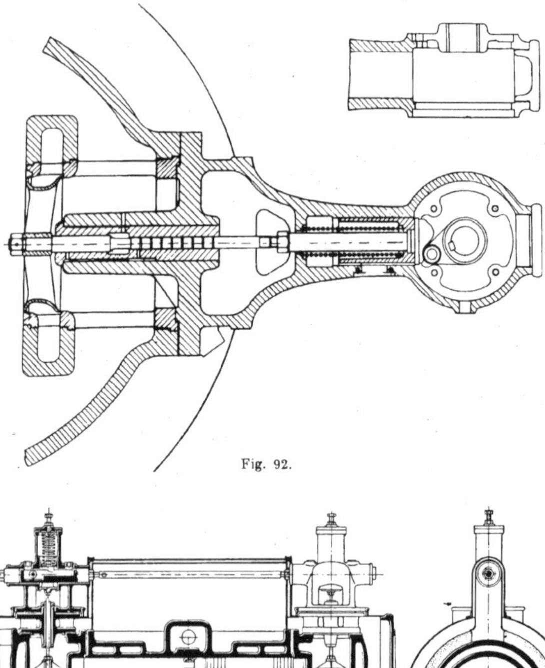

Fig. 92.

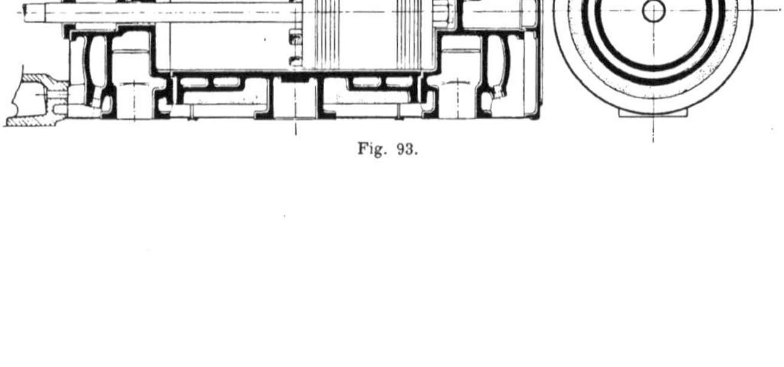

Fig. 93.

rod within the lagging of the cylinder, as shown in figure 93. With this construction the cylinder wall temperature is, to a considerable extent, communicated to the valve rod. In smaller engines it is always advisable to operate the valves directly from the main shaft, that is without any special valve operating shaft, as will be seen on examining the figures showing the engines built by Maschinenfabrik Grevenbroich and by Messrs. J. Musgrave and Son Ltd., Globe Iron Works, Bolton.

In figure 94, the details of the valve bonnet are shown. The cam piece is fixed to the sliding head on the valve stem, whilst the roller is arranged in a groove in the reciprocating valve rod. This groove also acts as an oil bath. The sleeve in the valve bonnet, in which the reciprocating valve rod is guided, is so extended towards the sides that the oil bath in the roller groove is never exposed. In this way all leakage of oil or ingress of dirt is prevented. It will be seen that the lubricant collects in the bottom of the oil bath and, as the roller rotates, oil is raised up on the surface of the roller to lubricate the cam, so that the cam and roller, which form the principal parts of the valve gear, are always well lubricated. It is probably due to this arrangement of the operating parts, that this form of valve gear works so reliably.

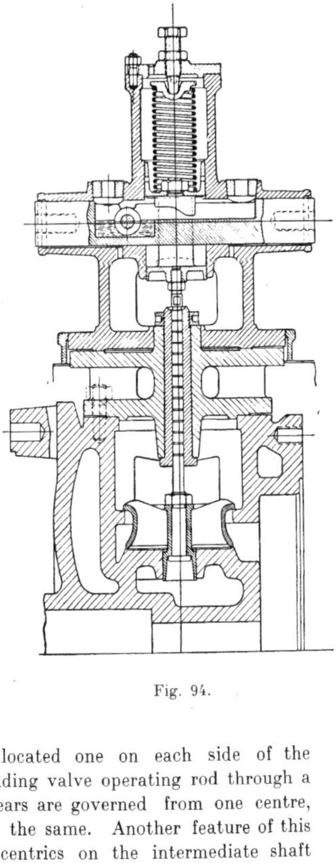

Fig. 94.

A twin una-flow steam engine is shown in figure 95, in which the roller and cam valve mechanism is operated from an intermediate shaft located between the two engines and driven from the main shaft. Two eccentrics on the main shaft operate two cranks on the intermediate shaft. The eccentrics on the main shaft and the cranks on the intermediate shaft are set at 90°. The intermediate shaft supports a shaft governor for controlling the two shifting eccentrics located one on each side of the governor. Each eccentric drives its corresponding valve operating rod through a rocking lever. In this way, both the valve gears are governed from one centre, and the cut-offs in both engines are always the same. Another feature of this valve gear is that the short vertical eccentrics on the intermediate shaft enable the cut-offs to be practically equal on both ends of the cylinders. The small intermediate shaft also gives greater freedom for the construction of the governor and its regulation to suit any given speed. This engine also gives a very

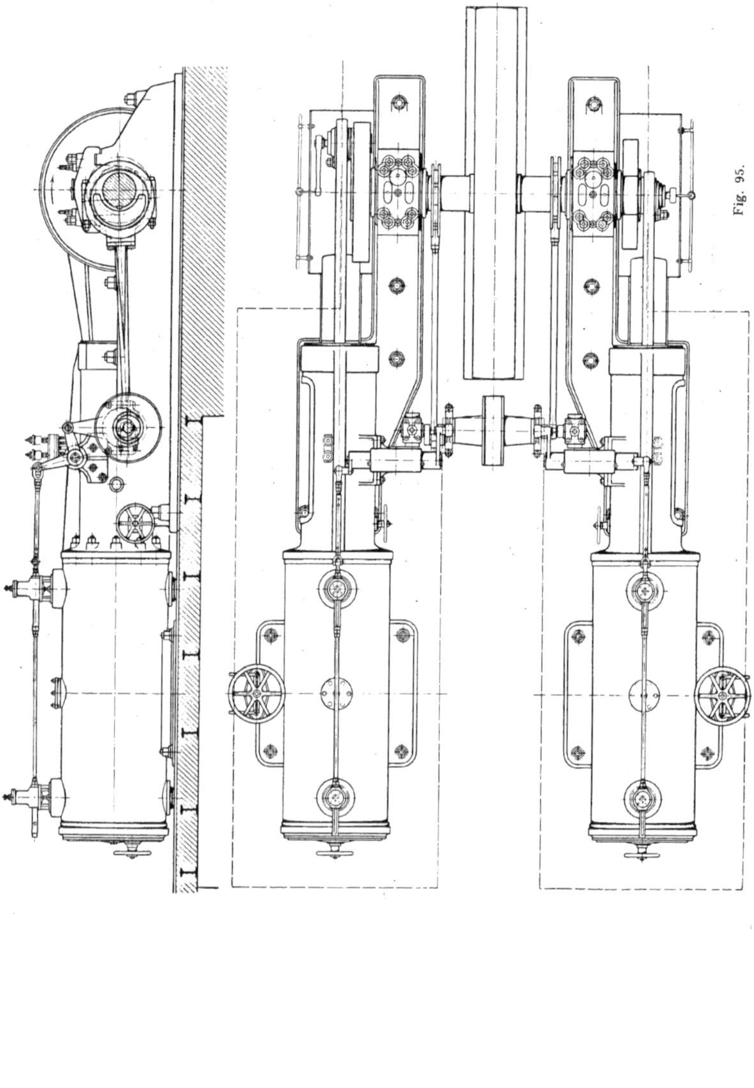

Fig. 95.

large reserve of power, as eachside is capable of taking up the whole load for a short time.

The una-flow engine may also be used as a vertical engine as shown in figure 96, which shows a 30 H.P. ship lighting engine built by Messrs. Frerichs & Co., Osterholz-Scharmbeck.

Fig. 96.

An interesting construction of a small vertical una-flow high speed engine, designed by the author for the firm List & Cie. in Moskau, is shown in figs. 97—101. The engine is intended to drive a direct connected continuous current dynamo. The horizonal single beat valves are made in one piece with the valve stem and are operated in the usual way by means of cams and rollers from an eccentric. The lower cover is cast in one piece with the cylinder, the cover jacket extending into the cylinder jacket. The upper cylinder jacket communicates with the upper cover jacket by means of openings arranged in the joint.

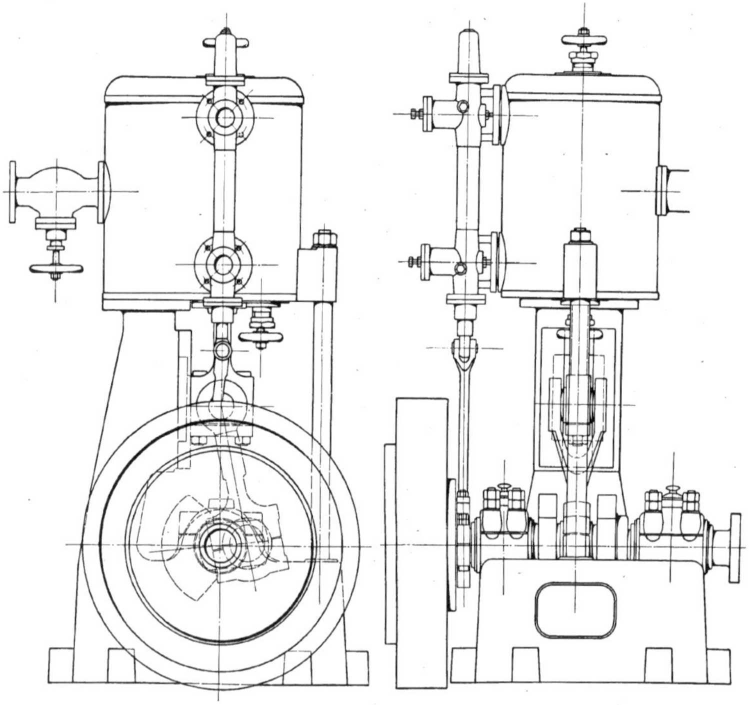

Fig. 97.

The live steam is introduced and divided into both end cylinder jackets by a breeches connection cast on the cylinder. The exhaust belt is separated from the end jackets by neutral unheated zones. The engine is intended to work with saturated or moderately superheated steam. The upper face of the piston is made conical, so as to favour scavenging of the water and effectful drying and heating afterwards. The exhaust holes are drilled by an electric driven drilling machine.

The additional clearance is arranged in the central part of the upper cover and at the side of the lower cover where it extends to the steel column in front.

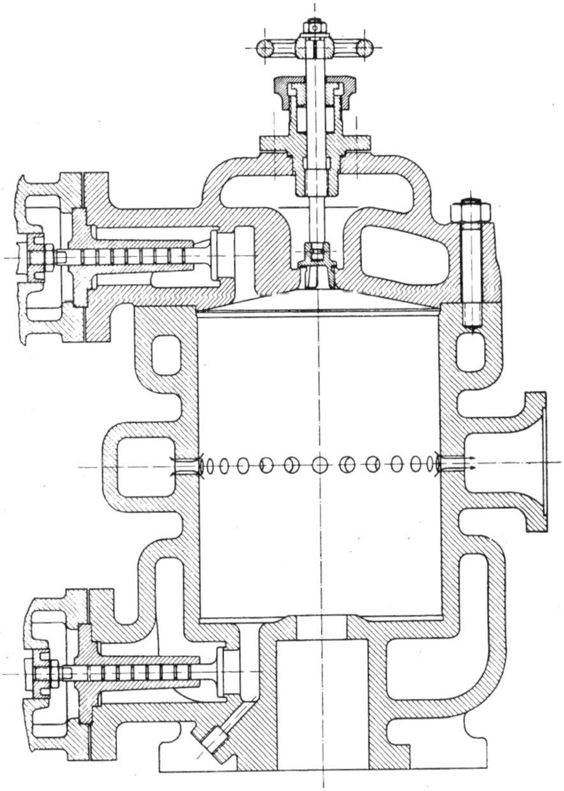

Fig. 98.

The design of the single acting twin cylinder una-flow engine, shown in figure 102, is especially worthy of note. In this design, due consideration has been given to all the requirements found from the experiments, as well as those found by care ful study of the conditions existing. The inlet valves are simple flat lift valves and are located centrally in the cover. These valves are operated from a governor-controlled eccentric on the main shaft, through the medium of a bell crank lever and a sliding rod, which carries the operating roller. The cranks are arranged at

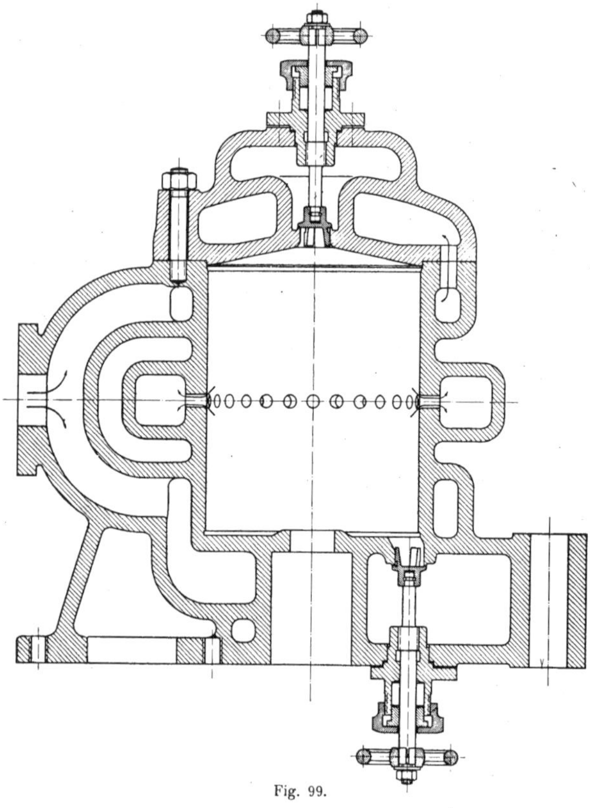

Fig. 99.

180⁰ to give good balancing at high speeds. The flat inlet valves are permissible on account of the high compression, which balances the steam pressure. The valve gear, however, is to be designed strong enough to work the valves properly even if there were no compensating compression pressure. The inner cover wall is made of thin

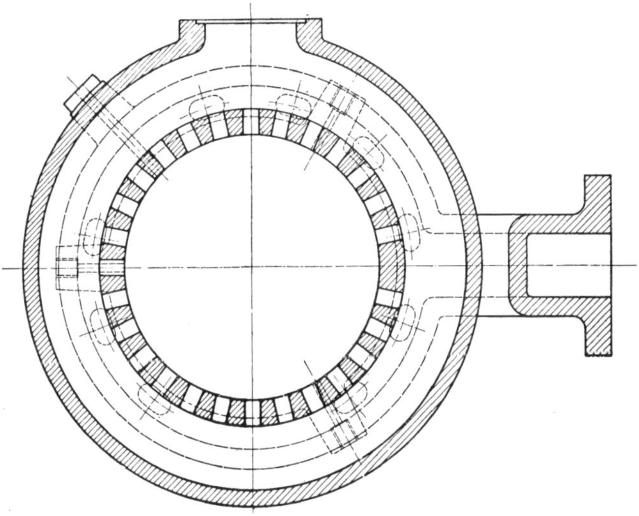

Fig. 100.

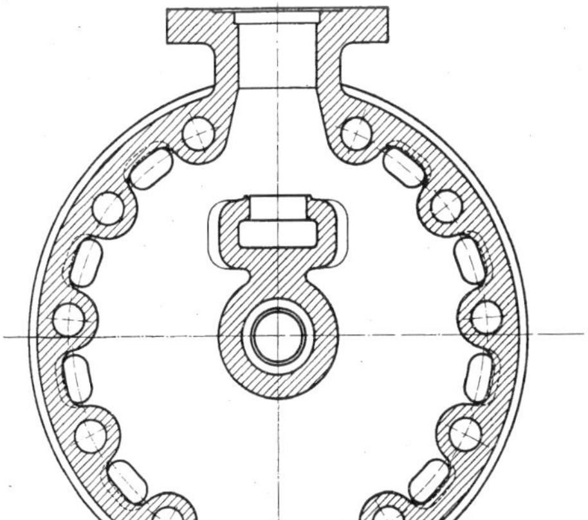

Fig. 101.

mild steel plate. A thin steel liner is also fitted in the cylinder. The cover is heated by the working steam and the cylinder is heated by steam introduced into grooves turned in the cylinder casting. These grooves are connected in series at diametrically opposite points so that the steam follows a zig-zag path from the top to the

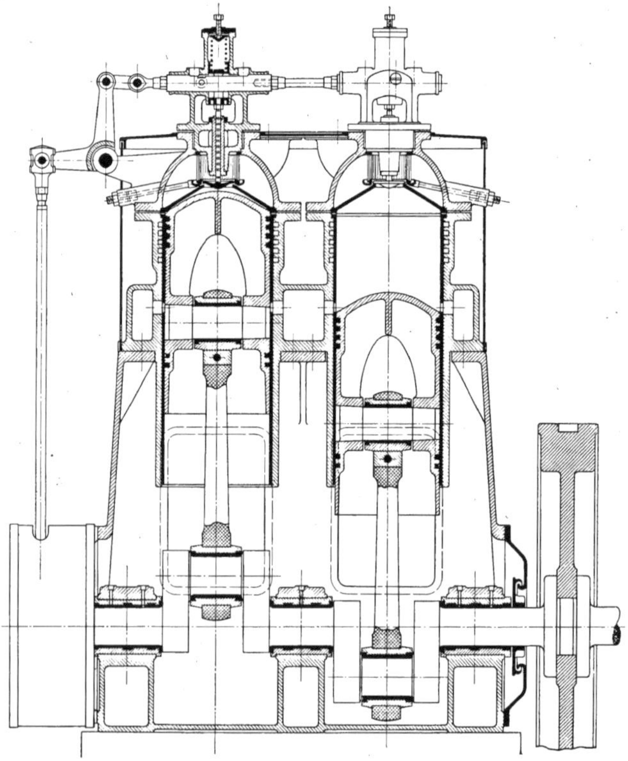

Fig. 102.

lowest groove. The engine is designed to work with saturated steam, hence the exceedingly short neutral or unjacketted part of the cylinder located between the heating grooves and the cold exhaust belt. The cylinder liner is preferably made of very hard material so as to work well with the cast iron piston rings. The end of the piston is convexed so as to enable the use of a conical inner plate on the cover, as with this form the required rigidity and strength may be obtained with very thin metal. This thin metal cover plate permits of a very rapid heat exchange from the hot end jacket to the internal clearance surface of the cylinder. Then again with this construction, the clearance surfaces are reduced to a minimum as was pointed out in the table given in Chapter V. The heat pumping action described above as produced by the compression, acts on the minimum amount of clearance surface, *so that with this construction the maximum heating action per unit of clearance surface is obtained* A water separator is arranged around the inlet valve so that the steam may enter the cylinder in as dry a state as possible. The conditions of steam flow through this engine

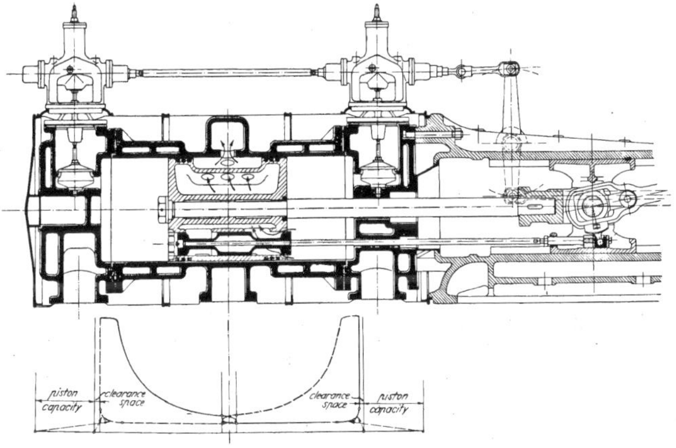

Fig. 103.

are more complete and perfect than will be found in any other. The steam enters centrally, when the piston is at the top of the cylinder, and distributes itself evenly towards the sides through the narrow conical space between the piston and cover. Then when the exhaust ports are uncovered by the piston, the steam sweeps over the piston surface downwards to the exhaust ports. The water, separated from the steam during expansion, naturally collects on the convex piston end, and this water is scavenged or swept down the inclined surfaces of the piston end and out to the exhaust. The surfaces are thus preliminarily dried and prepared for the more perfect drying and intense heating operations which take place during the compression. The working stroke always commences with considerable superheat. The separation of the water from the steam before the inlet to the cylinder and the scavenging of the moisture at the end of exhaust together with the consequent intense drying and heating action are features of great importance.

The steam consumption figures obtained with triple and quadruple expansion engines working with saturated steam may be excelled with an engine of this type.

The una-flow engine shown in figures 103—105 is provided with an additional exhaust valve in the piston and is specially intended for working with high counter pressures as for instance in cases where the exhaust steam is passed to a turbine or heating plant etc. In cases such as these, it is necessary to provide for extension of the period of exhaust so that a compression for about 5% of the stroke will occur. For this purpose, a valve, shown as a piston valve, is arranged in the piston and

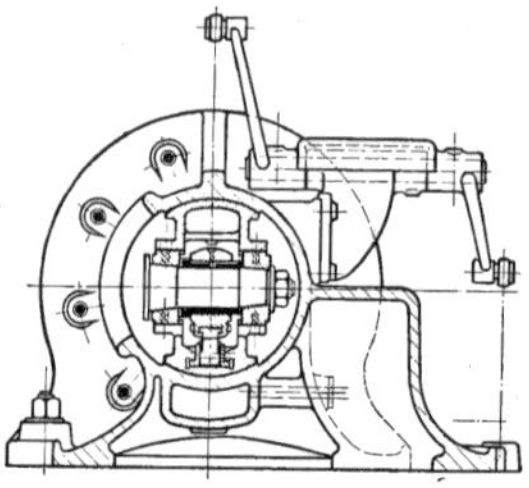

Fig. 104.

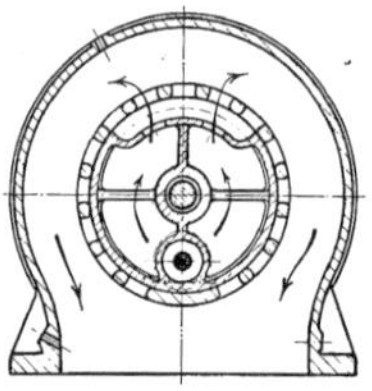

Fig. 105.

this valve is moved relatively to the piston by means of an arm on the cross head end of the connecting rod. In this way the exhaust ports in the cylinder are connected to the cylinder space even after the piston has overrun and covered these ports on its compression stroke. The residual steam in the cylinder can thus pass, by the exhaust valve in the piston, into the interior thereof and thence through openings in the piston wall to the exhaust ports in the cylinder. With this device, it is possible to extract the energy of the steam, even with high counter-pressures, entirely on the unidirectional flow principle and as fully as is possible with the best designs of multiple expansion engines.

It is particularly advantageous to operate the auxiliary exhaust valve in the piston from the crosshead end of the connecting rod as this gives a very sharp and rapid opening and closing of the valve in the neighbourhood of the dead centres, where the speed of the vertical oscillating movement of the rod is highest.

Chapter VII.

The una-flow engine, in combination with accessory steam-using apparatus.

With a una-flow steam engine cylinder, it is possible to withdraw working steam from the cylinder at intermediate points of its length by means of non-return valves as shown in figure 106. About $^1/_{10}$ th of the stroke before the exhaust ports

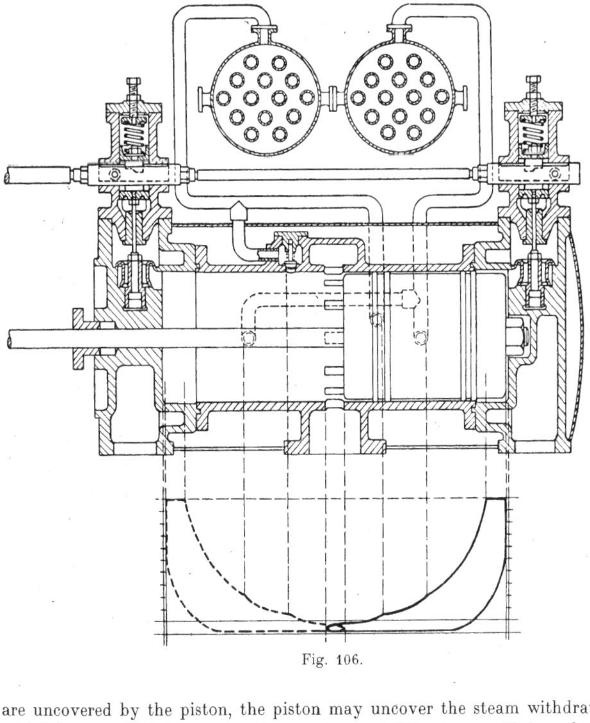

Fig. 106.

are uncovered by the piston, the piston may uncover the steam withdrawal port, which is controlled by a non-return valve and this steam withdrawal port may be connected to any accessory apparatus such as a heating system, a low pressure turbine or engine or any other steam using apparatus or machine.

It is possible in this way to withdraw steam of ¾ to one atmosphere absolute pressure from the cylinder of a condensing una-flow engine and to utilize this steam in a steam turbine for operating a rotary pump or the like for the condenser. If desired, the withdrawal ports can be arranged nearer the inlet end, and the withdrawal steam, which would then be of a higher pressure, could be led to some steam heating system. As shown in figure 106, several heaters may thus be arranged in series and these heaters may be used for bringing the boiler feed to a high temperature. This method naturally results in a small loss in the area of the

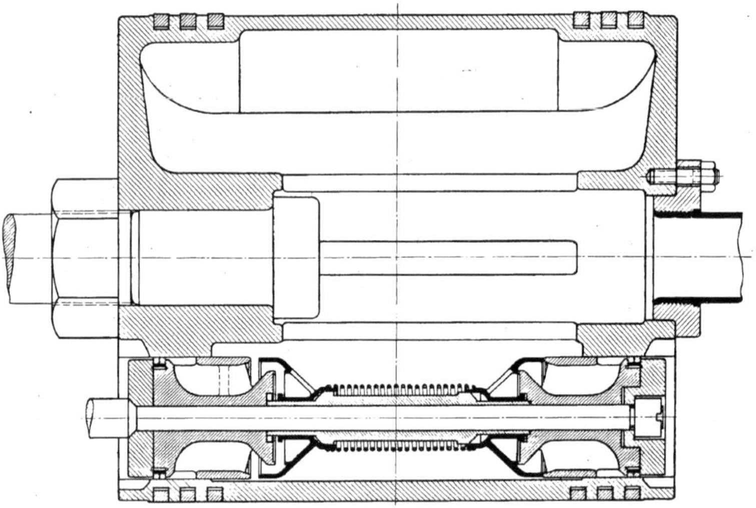

Fig. 107.

diagram. The quantity of steam withdrawn may be regulated by a governor to meet the needs of the accessory apparatus to which it is fed.

In figure 106 the working steam is withdrawn during the working stroke. It is also possible however, as shown in figure 107, to tap the steam from the cylinder during the compression stroke. The valve controlling the withdrawal of the steam, in such a case, is introduced in the exhaust passage, whereby the counterpressure and consequently the quantity of steam withdrawn may be increased up to the maximum amount of the steam working in the engine. By withdrawing the steam on the compression side, it is possible to meet extreme or widely varying demands from the heating apparatus. Figure 92, shows how a butterfly valve, arranged

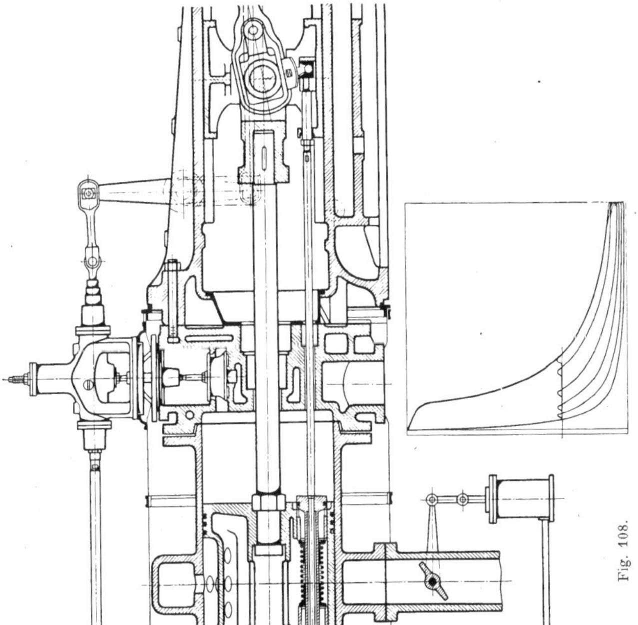

Fig. 108.

in the exhaust pipe and controlled by a pressure regulator, may throttle the exhaust and thereby regulate the quantity of steam withdrawn according to the requirements. The throttling of the exhaust in this way is not to be regarded as objectionable, as a sufficient expansion and the excellent thermal action due to the high compression must be added to the advantage of withdrawing any desired amount of steam. The steam withdrawn passes through a piston valve located in the working piston and also through a spring loaded valve in the piston. From the interior of the piston the steam passes through special ports in the piston itself to special exhaust ports, or from the hollow tail rod shown in figure 108, to the accessory apparatus.

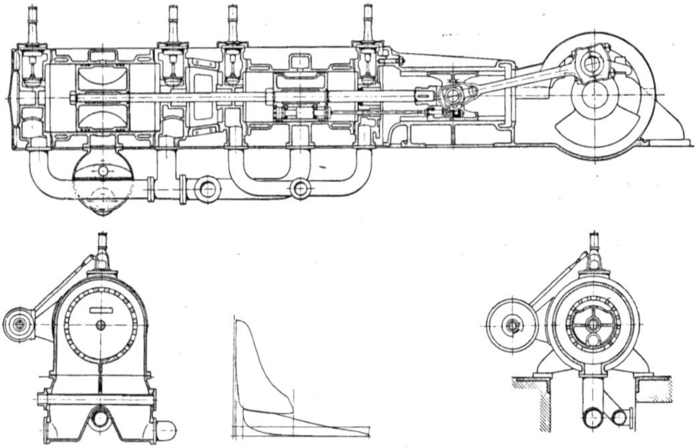

Fig. 109.

Modern developments have made it a very important problem to withdraw steam from a steam engine for use in low pressure apparatus and this problem may be solved very fully by using a form of compound una-flow steam engine such as the tandem arrangement illustrated in figure 109. The high pressure cylinder in this case is provided with an exhaust valve in the piston as described above, whereby the prolongation of the exhaust, necessary with high counter pressures, is obtained. The low pressure cylinder is of the ordinary una-flow construction. The clearance space in this L P.-cylinder should be increased to about 5—7%, in view of the low pressure in the covers (which also form the steam chests) of the L P.-cylinder. The volume of the intermediate receiver, that is the intermediate piping and the two end covers of the low pressures cylinder, should be retained as small as possible. The cut-off in the L. P. cylinder may be regulated by a pressure governor subjected to the pressure between the two cylinders. When the pressure between the cylinders falls, the cut-off in the L. P. cylinder is made earlier; when it rises, the cut off is made later. The L. P. cylinder in certain circumstances may be made the same size as, or even smaller than, the high pressure cylinder according to the amount of steam which will probably be required in the accessory apparatus. A large variety of excellent pressure regulating devices are known, which may be applied with advantage in an engine of the type just described.

Chapter VIII.
Una-flow locomotive engine.

A 4/4 coupled goods locomotive built by the Kolomnaer Maschinenbau A.G. in Kolomna, Moscow, for the Moscow-Kazan Railway, is shown in figure 110. This was the first una-flow locomotive furnished with una-flow steam engine cylinders designed by the Author and was built to the order of Mr. Noltein.

Fig. 110.

Fig. 111.

The engine was provided with a small auxiliary exhaust slide valve, which was dispensed with later, so that the engine now works as a pure una-flow engine. According to experiments, some of which were made with the auxiliary exhaust valve, this engine proved economical. It has been in constant use since it was introduced.

A 4/4 coupled goods locomotive working with superheated steam is shown in figure 111. This engine was built by the Stettiner Machinenbau Aktiengesellschaft "Vulcan" for use on the Prussian state railways. In figure 112, the locomotive is shown in longitudinal and cross section, and a longitudinal section through the cylinder of this engine is added. The simplicity of the cylinder

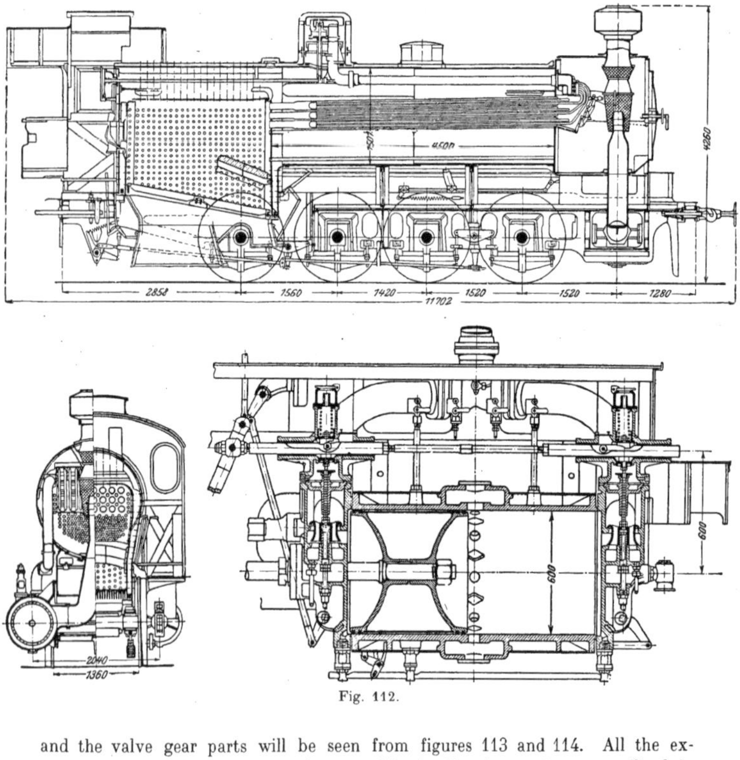

Fig. 112.

and the valve gear parts will be seen from figures 113 and 114. All the experience gleaned from the experiments with the Russian engine was utilised in this design. The engines proved remarkably successful, and since their introduction have been subjected to severe and steady night and day work.

The valves and steam chests are arranged in the covers. The exhaust ports and the exhaust chamber are in the cylinder casting. The features of this construction are: (I) The absolute separation of the hot working steam from the cold exhaust steam, (II) The introduction of the hot steam into the covers at a point where the high temperature has no deleterious action on cylinder and piston, (III) The arrangement of the exhaust space in the central part of the cylinder exercises a cooling action on the cylinder and the piston. These three features are all of the utmost importance, not only from a thermal point of view, but also from a mechanical working point of view. When the steam chest is cast in one with the cylinder, as is at present customary, the cylinder casting is complicated and must necessarily be subjected to considerable internal stresses, which result

Fig. 113.

in an unfavourable action of the piston, especially during the first few days of working.

A glance at figure 113, will show how simple is the construction of the unaflow cylinder as compared to the usual form of cylinder.

As the locomotive engine works with exhaust to the atmosphere, the clearance space must be considerable. In the case under consideration, when working with superheated steam at 12 atmos. pressure, the clearance was 17½%. This large clearance space is for the greater part accomodated in the concave end of the piston. With this form, the piston is of exceedingly strong construction as the ends of the piston are almost hemispherical. These piston ends are each provided with two piston rings and the hubs of the two ends extend towards one another but do not touch. These two ends are pressed together by the nut on the end of the piston rod and grip between them the steel cylinder which forms the outer closing wall of the hollow piston. This outer shell is made of hard forged steel containing 0·40% to 0·50% carbon. The thickness of the steel shell is on an average 6—7 mm. and its play is 0·003 of the cylinder diameter. Thus in the case of a cylinder of 1000 mm. diameter, the piston diameter is 997 mm. In this case the centre line of the piston rod lies 1½ mm. below the central axis of the cylinder. This allows for an expansion of the piston in excess of the cylinder.

The lubricating oil is not introduced into the steam pipe or steam chest, but is led direct to various points in the cylinder. Three lubricating holes are provided at each end of the cylinder, one at the top and two at points about 30° below

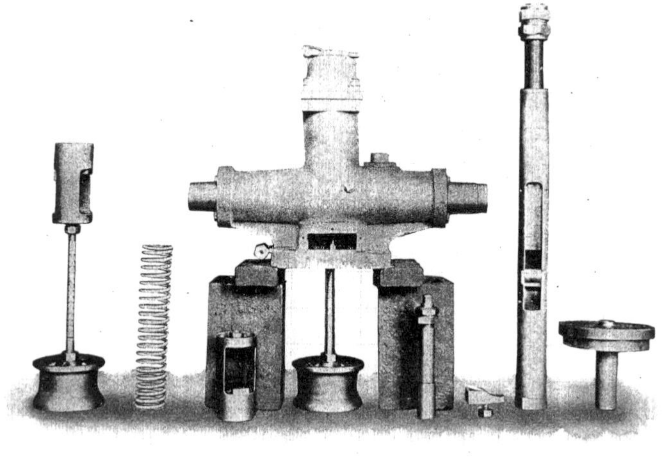

Fig. 114.

the horizontal. Six separate force feed lubricators arranged in one apparatus are provided, which supply oil separately to the different oiling places.

Owing to the concave ends on the built-up piston, the piston rod is shortened and the large bearing surface of the piston renders a tail rod and a stuffing-box for same superfluous. The increased weight of the long piston (which is 9/10 stroke in length), is to be written against the reduction in the weight of the piston rod. The weight of the una-flow piston does not exceed, however, the weight of the L. P. piston of a compound engine of the same power.

The volume of the clearance space is dependent upon the pressure and temperature of the admission steam. Figure 115 shows a number of compression curves for different pressures and temperatures. The compression diagrams show the necessary clearance spaces, assuming the use of perfectly steam tight valves (as described in Chapter IV), compression for 90% of the stroke, and a final compression pressure which is 2½ atmos. under the initial pressure. The working steam is assumed as saturated, (dryness fraction, $x = 0{\cdot}97$). This dryness fraction for the working steam gives, with adiabatic expansion, the various values marked on the diagram for the dryness fraction at the commencement of compression. In

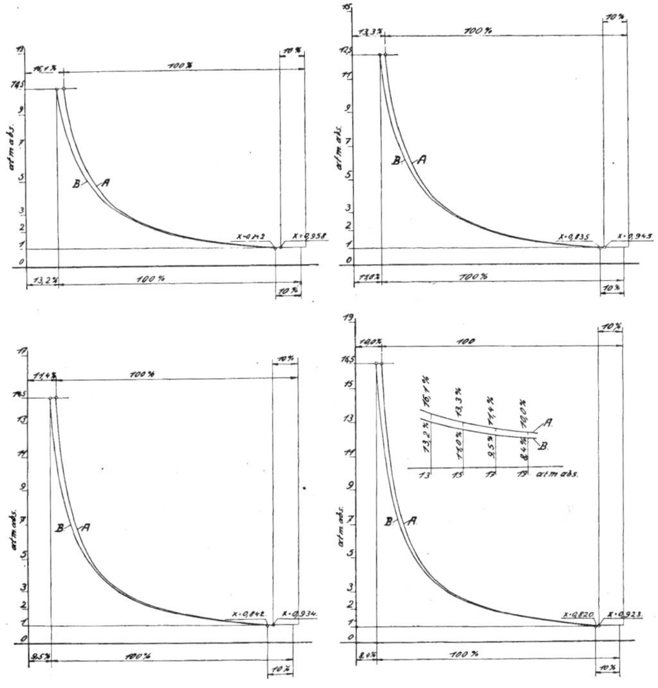

Fig. 115.

addition, there are added the corresponding dryness fractions at the commencement of compression when the working steam is superheated to 325° C. The resulting clearance volumes are given in the last diagram as functions of the initial admission pressure. The curve marked *A* is for superheated steam, whilst that marked *B* is for dry saturated steam.

The final diagram above gives with 16 atmos. admission pressure, a 9·5% clearance for saturated steam and a 11·4% clearance for superheated steam. With working steam at 18 atmos., the clearance for saturated steam is 8·4% and for superheated steam 10%. These values are below the usual clearances in ordinary modern constructions.

In Chapter III tests of a una-flow engine were given, which seem to point to the conclusion that superheating is unnecessary. If the same reasoning is applied to the una-flow locomotive engine, it will be found to lead to a similar conclusion. Superheating, in some cases, may be advantageous in a unaflow engine, when the temperature drop is very great. The advantage is smaller, however, in proportion to the reduction in the temperature fall. The temperature fall in a non-condensing engine is considerably less in comparison with a condensing engine. To this consideration has to be added the fact that high superheats are liable to increase the losses by radiation and leakage. The increased leakage is due to the internal stresses in all steam distributing parts. The superheat in locomotives varies considerably with varying load, it rises with increase of load and falls with reduced load. The thermal requirements of a steam cylinder are just the converse of this, viz: at early cut-offs the superheat should be high and at late cut-offs, low. That is to say, the superheater works in exactly the converse manner to that in which it should work to meet the requirements of the engine. At high loads, the exhaust steam passing to the chimney contains a considerable degree of superheat. Temperatures as high as 200° C have been registered in the exhaust pipe. This heat is entirely lost. Then again the pressure loss due to the fluid friction in the superheater is about 1—1·5 atmos.

Possible lines of development would be to employ saturated steam, introduced into the cylinder in as dry a state as possible, or superheated only a few degrees and at a pressure that is at present usual in compound locomotives, so that the clearance space and the loss entailed thereby, may be reduced.

A una-flow locomotive engine working with high pressure saturated steam up to say 16—18 atmos., should prove not only a very economical engine but should also be simple and cheap in construction and more adapted to the working requirements of a modern locomotive.

It would be possible to increase the boiler pressure in modern locomotives by about 2 atmos. without increasing the load on the axles if the superheater were dispensed with and the weight of the superheater made up by thickening the boiler plates. This gain in pressure combined with the fluid friction loss in the superheater gives a total gain of 3 atmos. in favour of the engine working with saturated steam.

On behalf of superheating, it must be admitted that it offers the great advantage of providing a dry working medium and a considerable increase in the volume of the steam, as well as avoiding initial condensation and thereby permitting of the use of a larger cylinder and greater tractive power. Probably the most prominent feature in favour of modern superheater locomotives lies in the increased tractive force. The tests made on una-flow steam engines show that this type of engine gives excellent thermal efficiencies with saturated steam, so that the cylinder may be increased and consequently the tractive force. This statement is fully borne out by the tests made on existing una-flow locomotive engines, which show the advisability of making the dimensions of the cylinder in excess of those usually employed in counter-flow locomotive steam engines. The covers and the ends of the cylinder should be jacketted when using saturated steam.

The question now arises, is it preferable to employ the counter-flow or the una-flow system with high temperature steam? It must be answered that even with high temperature steam the advantage lies with the una-flow engine. Although the disadvantages present or consequent upon the use of the counter-flow system are considerably counteracted when the entire cycle is completed at a temperature above the saturation point, there is still a loss which is not removed or corrected by superheating. This is especially the case when running in the ordinary manner with early cut-off, when, unfortunately, the degree of superheat is low and therefore the cycle of operations in the engine is at such times carried below the saturation point. Under these circumstances, the large cylinders, rendered possible by the introduction of superheating and favoured on account of the increased tractive power they yield, are not advantageous.

The loss of heat from the clearance surfaces, entailed by the counter-flow action, is present also in this case, although in a lesser degree, on account of the superheat. With an initial temperature of 270° C, and with expansion from 11 atmos. absolute to one atmos. absolute, the saturation point is reached with adiabatic expansion when the pressure is 3·25 atmos. absolute, and when the pressure falls to atmospheric, the steam contains 6% moisture. It follows from this, that in the ordinary counter-flow locomotive engine working with an early cut-off and superheated steam of 270° C, the steam temperature at the end of expansion has already fallen below the saturation point and contains some moisture. This moist steam abstracts heat from the clearance surfaces which have been heated by the incoming steam during admission and the exhaust steam passes off with considerable superheat. This may be proved by referring to figure 35, in which it is seen that the exhaust steam is superheated at the very commencement of the outflow and the superheat rises considerably towards the end of the exhaust. In this case the working steam was saturated, but the cylinder was heated by means of steam at a higher pressure than the working steam.

The most unfavourable feature of the counter-flow engine is the fact that the admission steam and exhaust steam pass through the same ports, and this very materially increases the amount of the admission heat which is carried off by the exhaust. The una-flow engine acts in exactly the opposite manner and consequently under much more favourable conditions, the cold exhaust steam being withdrawn away from the inlet end to the exhaust ports. It has been proved when working with steam at 300° C, that the exhaust was actually wet steam when the cut-off was early. The inlet valves and piston must necessarily in such cases be absolutely steam tight.

Comparative tests with una-flow and counter-flow engines working with superheated steam of 11 atmos. have shown that the advantage rests with the una-flow engine for light and medium loads, but for heavy loads and overloads the advantage rests with the counter-flow engine. This advantage of the counter flow engine over the una-flow engine may be materially reduced, and the range of loads, with which preference must be given to the una-flow engine, materially increased by employing a larger cylinder for the una-flow engine. Such increase in the cylinder volume is rendered possible on account of the favourable thermal conditions under which the engine is working. By increasing the cylinder volume, the tractive power is also increased. The output per stroke of the engine may also be increased by using a compression reducing valve as shown in figure 120.

Other means of improving the action of the una-flow engine for loco work are to increase the boiler pressures and consequently the compression, so that the clearance volume may be reduced as described with reference to figure 115. This reduction of the clearance volume increases the expansion ratio and would probably make the una-flow engine equal to the counter flow loco engine at overloads, while the advantage at light and medium loads would still rest unquestionably with the una-flow engine.

A most important feature of the una-flow loco engine is the automatic scavenging action introduced. Dirt, grit, scale or anything of the kind carried over by the steam is also caused to follow the uni-directional flow, so that anything passing the inlet valve is swept out at the exhaust ports into the exhaust belt out of which it may be withdrawn by an opening at the bottom of the belt. This is the only possible reason for the excellent condition of the piston and the interior of the cylinder of the una-flow loco engine when these are examined after running for some time under the most severe working conditions. The dirt and grit scavenged from the interior of the una-flow engine, remain and collect for the greater part in the ordinary engine cylinder, from which they are only probably partly removed by opening the sludge or drain cocks, and even then these drain cocks are only open from time to time. The scavenging action in the case of the una-flow engine is constantly at work or is effective after each stroke, so that there is only a small amount of dirt to be scavenged out and the action is therefore more reliable. The dirt never seems to stay in the cylinder long enough to do any damage, hence the splendid condition of the cylinder, piston and rings, which is another point in favour of the una-flow engine. This preservation of the inner surfaces, not only reduces friction, but also reduces leakage losses.

The hole in the bottom of the exhaust belt also allows the water of condensation to drain off. In the loco engine, the working steam passes down from the boiler to the engine and then the exhaust passes up from the engine to the chimney. The cylinder is thus located at the lowest point in the path of the steam where it is well situated to act as a water pocket. Hence the danger of breakdowns caused by water in the cylinder. This is quite impossible with the una-flow engine, where the annular exhaust belt, being at the lowest point, acts as a separator and water collector. All water drains through the exhaust ports into the lower part of the exhaust belt, from which it passes off through the small opening mentioned above.

Another feature of the una-flow loco engine is the "coursing" mechanism. This consists of cams, which are operated by hand from the driver's cab and which may raise the inlet valves and hold them raised with their cams out of contact with the roller on the operating rod. In this way the two cylinder ends are connected by the legs of the breeches pipe from the regulator valve, and the steam merely pendulates to and fro through this pipe from end to end of the cylinder. It is preferable to raise the inlet valve by a considerable amount, and in excess of its normal full lift, so as to ensure that the valve cam will not contact with the roller, and also to prevent the temperature of the pendulating steam being so increased by friction that the lubricating oil is carbonised. The una-flow locomotive engine thus lends itself excellently to coursing without employing excessively large supplementary clearance surfaces or spaces.

Exhaust commences at about 9/10 ths. of the stroke and ends at about 1/10 th. of the return stroke. When the piston uncovers the large area of exhaust passage

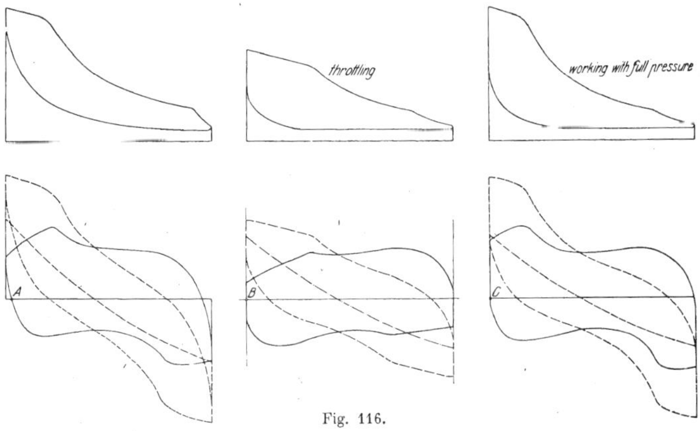

Fig. 116.

there is a sudden sharply defined blast, which improves the furnace draught. The exhaust blast is not extended over such a long period as in the ordinary counter-flow locomotive, but, by properly designing the blast pipe this defect may be removed to a considerable extent. For ordinary sized coal, the furnace blast obtained by a una-flow engine is better than that obtained by the ordinary locomotive. Any coal which has been raised from the fire falls back in the intervals between the exhaust puffs.

With light coal or dross, this may not happen and the coal might be carried out at the chimney. To avoid this, it is advisable to introduce a silencing box between the exhaust belt and the blast pipe.

When running light with an ordinary slide valve locomotive, the valve travel is small and the exhaust is throttled. This throttling of the exhaust is entirely absent in the case of the una-flow engine, as the exhaust is controlled by the piston which always uncovers the same large area of exhaust port. This feature obviates the losses, produced in ordinary loco engines at light loads by the high counter pressures.

It was demonstrated in the case of two una-flow express locomotives built by the Maschinenbau Anstalt Breslau to the order of the Prussian railways (see figure 124), that such engines run smoothly.

This smooth running is due to the method of steam distribution. In order to show this clearer, the diagrams from a una-flow locomotive engine are shown alongside the diagrams from a counter flow engine in figure 116. In the case of

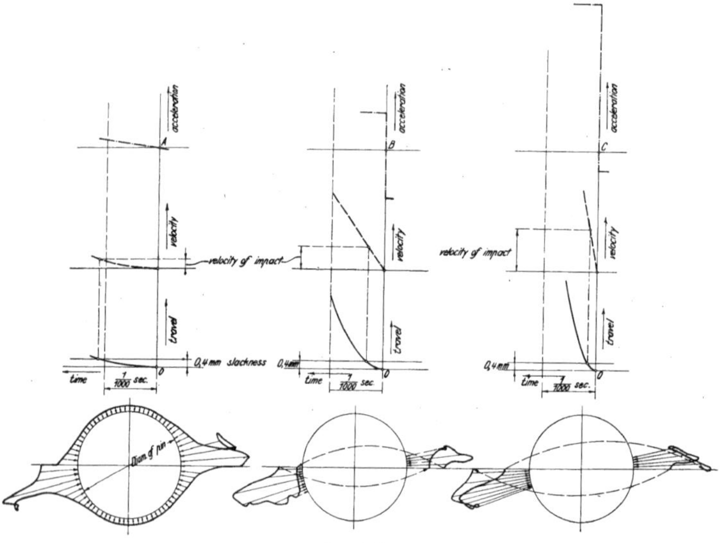

Fig. 117.

the counter flow engine it is shown working with throttling and with full pressure. In the full pressure diagram, cut-off takes place after 17% of the stroke, whilst in the case of the throttle diagram the cut-off takes place after 31½%. The cut-off for the una-flow engine is at 20% of the stroke. The inertia diagrams have been drawn for a speed of 110 km. and the rate of change of the pressure on the crank pin has been derived from the combination of the steam and inertia diagrams (Fig. 116, lower diagrams). The three upper diagrams, figure 117, show the rate of change of acceleration resulting from the effective pressure as a function of the time. It will be noticed that the curve of acceleration for the una-flow engine is very nearly level, whilst in the case of the counter flow engine there is a sharp change from positive to negative. This is due to the fact that in the counter flow engine, the change in the direction of the effort coincides with the dead centre, whilst in the case of the una-flow engine the pressure changes a considerable distance before the dead centre, on account of the high compression, and this result is obtained even with the comparatively high velocity mentioned. The somewhat greater weight of the piston of the una-flow locomotive engine has been taken into account. From the acceleration curves, are obtained the corresponding velocity curves and from these the corresponding space curves may be derived by integration. If a total play of 0·4 mm. for all pins is assumed, the horizontal line representing this play will intersect the space curves s, at the points where the shock occurs. By projecting these points up to the middle velocity curves v, the velocity at the time of the shock will be determined. The curves show that the velocity of the shock in the case of the una-flow engine is very much less than the velocity in the case of a counter flow engine with throttling, and still less than in the case of a counter flow engine working with full pressure of steam. The actual ratios in the cases under consideration are 1 : 2·55 : 4·7. If the weight of the reciprocating masses is taken into account it is found that the forces of the shocks are in the ratio of 1 : 6·35 : 21·55.

The above figures are very instructive in showing the reason why the crank pin of a four cylinder compound locomotive frequently fires, the reason being that the steam diagram pressure is low and the inertia diagram pressure high on account of the heavy piston, whilst the changing over from the positive to the negative effort takes place at the dead centre. The above investigation also shows the disadvantageous effect of the variation in compression. In the ordinary form of engine cylinder with a counter flow action, the change in the effective pressure always takes place at the dead centre, when the cut-off is late and the compression is small.

The three lowest diagrams in figure 117 represent in radial form, the forces acting on the crank pin. In drawing these diagrams the centrifugal forces coming into play and the effect of the finite connecting rod have been taken into account. The diagram on the extreme left shows the pressures on the crank pin in the case of a una-flow engine, whilst the two on the right show those of a counter flow engine. It will be seen that the pressure in the una-flow engine follows smooth curves, whilst that in the case of the counter flow engine is jerky and abrupt. There is, what might be called a gliding of the pressure as shown in the case of the una-flow engine, whilst the shock in the case of the counter flow locomotive is hard and metallic. It is interesting to note here how correct are the regulations of the Prussian railway authorities regarding notching up. The authorities forbid the linking up below 20% and require the locomotive to be regulated to meet the load on it thereafter by means of the regulator. The centre diagram shows that throttling gives a considerable reduction in the force of the shock. As determined from the diagram, throttling reduces the force or the work lost by the shock in the approximate ratio 22 : 6·5. The locomotive engineers have found these things out in practice and show a marked preference for throttling. The author has made many trips on locomotives and has frequently had opportunity of proving in practice the correctness of the above theoretical investigation.

By throttling, it is possible to improve the mechanical working of a counter flow locomotive engine to a remarkable degree. The same end might be obtained without throttling, if it were possible to make cut-off earlier and increase the compression to such an extent that the change in the direction of the crank pin pressure is clear of the dead centre and the angle with which the effective pressure line meets the inertia line correspondingly reduced. No useful purpose would be served by working a una-flow locomotive with throttling. In the case of a una-flow locomotive, the regulator should be full open and the engine governed from the link motion entirely.

The una-flow locomotive shows a diagram very similar to that of a Diesel-motor and in common with this motor runs exceedingly smoothly. The fundamental reason for this is to be found in the compression space common to both. By means of a suitable compression space in combination with a piston controlled exhaust, it is always possible to obtain the same high compression and good cushioning of the moving masses as well as proper timing of the pressure change on the crank pin and smooth running of the engine. *The above proves that the una-flow steam locomotive is especially suited for high speed express work.*

All the thermal advantages peculiar to the uni-directional flow system are present in addition to the above. These thermal advantages, in combination with the exhaust blast, which produces an exceedingly good draught, with a very small vacuum in the smoke box, account for the suprising results obtained as regards coal consumption with this new form of locomotive.

The Prussian railway authorities had a two months comparative test made with two una-flow locomotives, two piston valve locomotives and two lift valve locomotives. All three types of locomotives were provided with superheaters and were 4/4 coupled. In normal working they travelled over the same line. The engines were working with two shifts day and night and the conditions in all cases were as equal as possible, as also were the loads. The piston valves of the piston valve locomotive were not provided with rings and accordingly were leaky. The same might be said of the una-flow locomotive which was provided with the deep rigid valves shown in figure 114.

The coal consumption in the case of the una-flow engine was far and away less than that of the others engaged in the test. Since these tests, the results obtained by the piston valve locomotive were considerably improved by using piston valve packing rings. The same advantageous results however, were introduced with the una-flow engine by using resilient valves and by jacketting the covers and ends of the cylinder. Also data regarding the necessary dimensions of the cylinders were secured. The same remarks apply to the case of the locomotive having inlet and exhaust lift valves. But as yet no exact tests have been made with these three types of locomotives thus improved.

The diagrams shown in figure 118, illustrate the excellent conditions under which the steam is working in the cylinder. *All counter pressure is absent.* The una-flow locomotive came up to all requirements and in fact the expectations were exceeded in many respects.

The resistance of the valve gear is a minimum and it is not necessary to lock the gear in reversing position. The reactionary forces are so small that any movement of the linkage, when the lock is not in the notch, is absolutely impossible. No material wear of the parts of the valve gear occurs.

A resilient valve for an express locomotive engine together with its chair is shown in figure 119.

The valves are operated by a roller and cam gear, designed by the Author. The roller is carried in a groove on the horizontal reciprocating valve rod, this groove acting as an oil bath. The guiding sleeve in the valve bonnet is so long

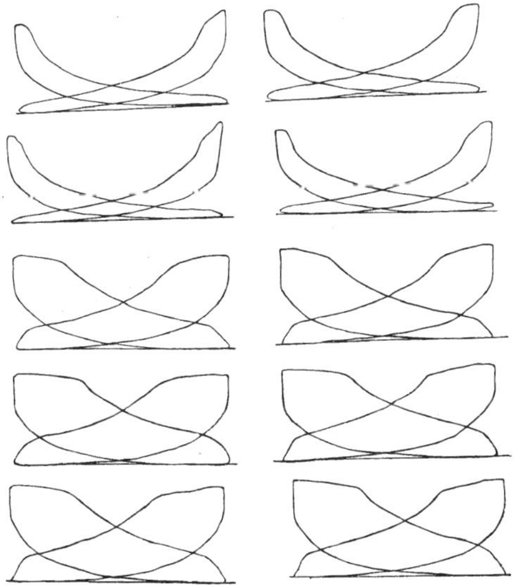

Fig. 118.

that this oil bath is never exposed to the outer air. The oil from the vertical guide head on the valve spindle collects, for the greater part, in the oil bath on the valve rod. The roller dips in the oil and carries it over to the cam, which is fixed to the guide head on the valve spindle, so that these important parts are always well lubricated. It is due to this efficient lubrication that no difficulty whatever has been encountered in this valve operating gear. The guide head on the valve spindle is lubricated by means of a wick. An ordinary Walchaert gear is employed without alteration for operating the valve rod. This gear was adopted from existing locomotives. Both roller valve rods may be adjusted by means of a thread connecting piece.

The valve spring is preferably adjustable and ought to be sufficiently strong to keep the cam and roller, without any unnecessary load, always in contact even at late cut-offs and at high speeds. This end is partly attained by employing a light forged steel valve which, on account of its small mass, only requires a light spring.

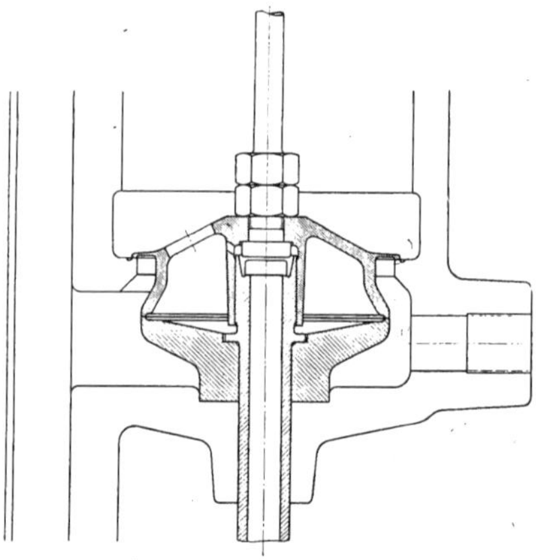

Fig. 119.

Figure 120, shows a double seated relief valve connected by two pipes to the ends of the cylinder. Two stop valves are inserted in these pipes and these valves may be operated by hand from the engine cab. When these valves are opened, the live steam admitted on one side presses on the corresponding side of the double valve shown at the centre and the bottom of figure 120, and holds the opposite valve plate open. In this way, the compressed steam may pass out on the compression side into the exhaust belt through the ports leading from the relief valve casing to the exhaust belt. The relief valve is shown in position to relieve compression on the right hand side of the piston. When the live steam is admitted on the opposite side, the relief valve will move from right to left, and the left hand valve plate will then be opened to allow the compressed steam to pass off from the left hand side of the piston. With this apparatus, high com-

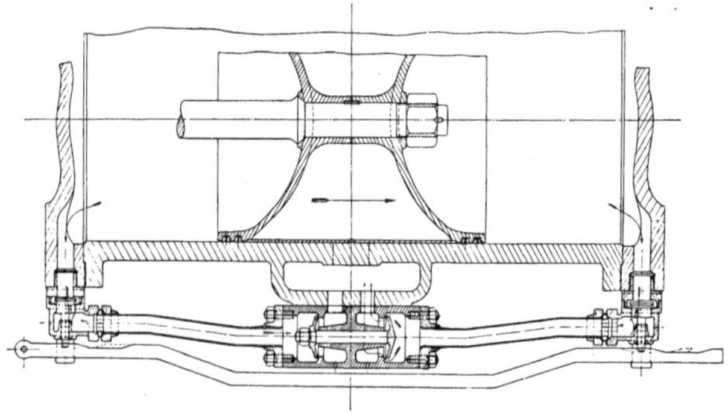

Fig. 120.

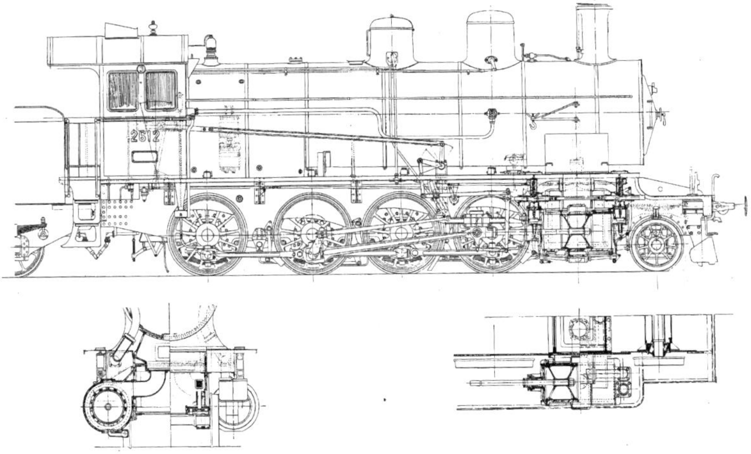

Fig. 123.

Printed by R. Oldenbourg, Munich and Berlin.

pression may be avoided when starting the locomotive, at which time of course the load on the engine is greatest. As soon as the engine is in motion, the two stop valves located near the ends of the cylinders in the connecting pipes to the relief valve may be closed. The operation of the engine then proceeds in the regular manner of the una-flow engine with high compression. The above described compression reducer is specially useful for express engines, which are frequently called upon to start a long train of carriages on a curved stretch of line.

Figure 121 shows in picture form a 4/4 coupled una-flow goods locomotive with superheated steam, the principal sections of which were shown in figure 112. This engine was exhibited at the International Exhibition in Brussels.

In view of the increased cylinder weight, the boiler is displaced a little to the rear of the bogie so as to give a better distribution of the weight on the axles. The principal dimensions of this engine are as follows:

Cylinder diameter	600 mm.
Piston Stroke	660 mm.
Diameter of the wheels	1350 mm.
Steam pressure	12 atmos.
Heating surface of boiler	140·42 sq. m.
Heating surface of the superheater . . .	38·97 sq. m.
Grate area	2·35 sq. m.
Weight (empty)	52125 kg.
Weight (service)	57750 kg.
Weight (effective gripping on rails) . .	57750 kg.

Figure 122 gives a picture of a 4/5 coupled una-flow goods locomotive working with superheated steam. Two such locomotives were built to the order of the Swiss Confederate Railway authorities by the Swiss Locomotive works in Winterthur. The dimensions of these locomotives are as follows:

Cylinder diameter	570 mm.
Stroke	640 mm.
Driving wheel diameter	1330 mm.
Diameter of the bogie wheels	850 mm.
Steam pressure	12 atmos.
Heating surface on the boiler	143·7 sq. m.
Heating surface of the superheater . . .	37·6 sq. m.
Grate area	2·44 sq. m.
Weight (empty)	60875 kg.
Weight (service)	6770 kg.

A longitudinal section and a plan of this locomotive is shown in Figure 123 (Plate 1). The cylinder is of the same design as described above. Between the exhaust belt and the blast pipe a silencer is inserted and this silencer is formed in the front saddle piece which supports the boiler.

A silencer of this kind is only advisable when using small coals, which are apt to be carried up through the funnel. With ordinary hard coal, the silencing

Fig. 121.

Fig. 122.

Fig. 124.

box may be dispensed with. Tests have shown that with a sufficient diameter of exhaust pipe and properly shaped exhaust ports, the noise of exhaust in the case of the una-flow engine may easily be reduced so as not to exceed that common in ordinary locomotives.

For this reason, in the case of the repeat orders received for una-flow locomotives from the Prussian State Railway Authorities, no silencer was used. Figure 124, shows a una-flow locomotive engine for superheated steam of a design from which two were built by the Maschinenbau Anstalt Breslau to the order of the Prussian State Railway Authorities. A third locomotive of the same kind

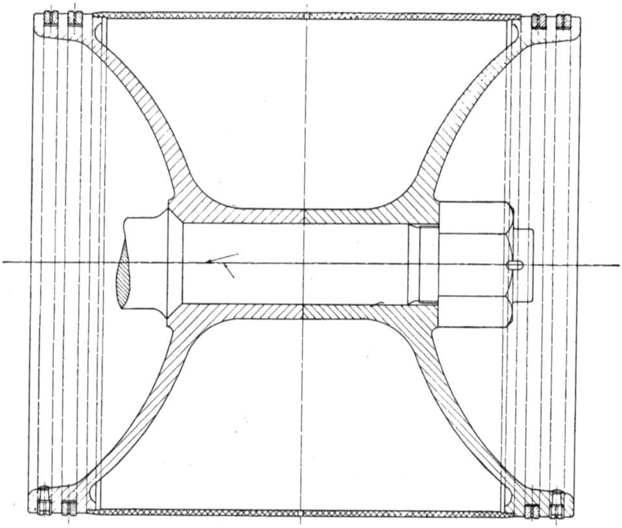

Fig. 125.

was built by this firm for the exhibition at Turin, and has also been taken over by the Prussian State Railway Authorities.

The principal dimensions of the engine exhibited at the Turin Exhibition are as follows:

Cylinder diameter	550 mm.
Piston Stroke	630 mm.
Driving wheel diameter	2100 mm.
Bogie wheel diameter	1000 mm.
Steam pressure	12 atmos.
Heating surface of the boiler	136·98 sq. m.

Heating surface of the Superheater. . .	40·32 sq. m.
Grate Area.	2·31 sq. m.
Weight (empty)	56 900 kg.
Weight (in service)	62 500 kg.
Weight (effective gripping on rails) . .	35 000 kg.

$$\text{Maximum tractive power } 0{\cdot}75\ p \cdot \frac{d^2 \cdot h}{D} = 8200 \text{ kg.}$$

Fig. 125 shows the piston, especially designed for superheated steam. Fig. 126 shows a piston, especially designed for saturated steam. Attention might be drawn

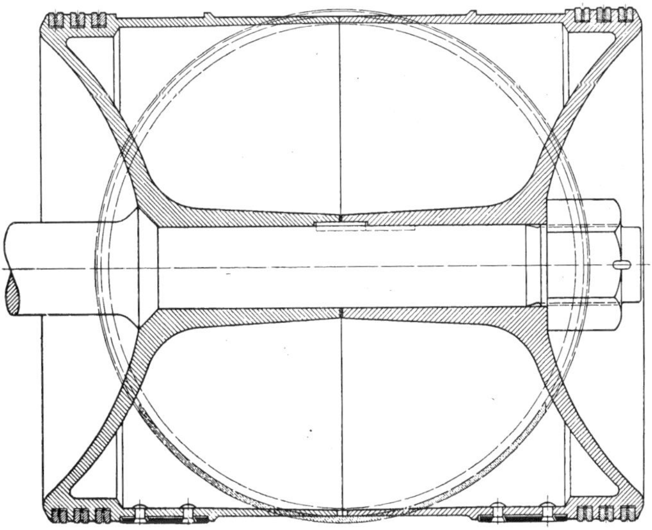

Fig. 126.

particularly to the brass armature employed in the construction shown in figure 126. Such an armature cannot very well be used whith superheated steam.

Figures 127 and 128, give the valve, valve bonnet, and roller cam mechanism of these locomotives.

The method, described in Chapter VII, of withdrawing steam from the cylinder is specially important for express locomotives where the steam withdrawn may be used for heating the carriages and also, if desired, for heating the feed water.

Figure 129 shows an arrangement of una-flow locomotive steam engine cylinder with automatic steam withdrawal valves. The diagrams shown in figures

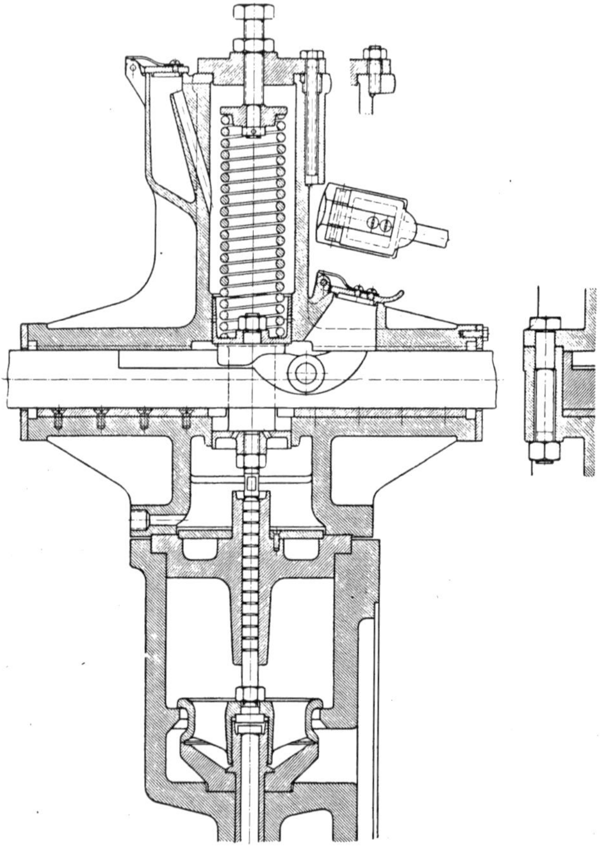

Fig. 127.

130 and 131 are for different speeds and illustrate the effect on the diagram when the automatic steam withdrawal valves are placed at 35 % and 50 % respectively before the end of the stroke. The corresponding quantities of steam withdrawn are given on the diagram. The amount of steam withdrawn increases with the area of the steam withdrawal valve and the distance of the valves from

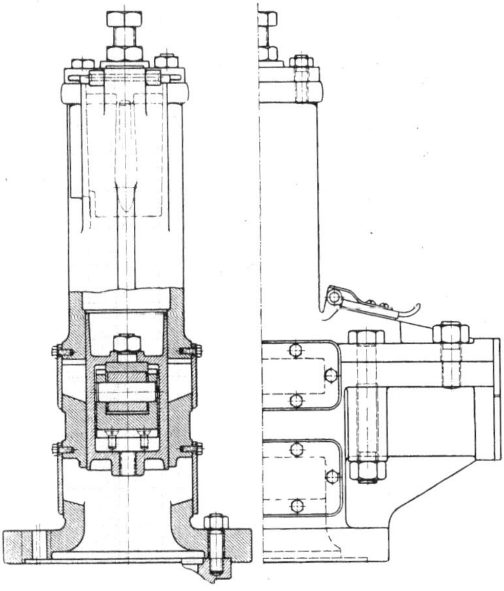

Fig. 128.

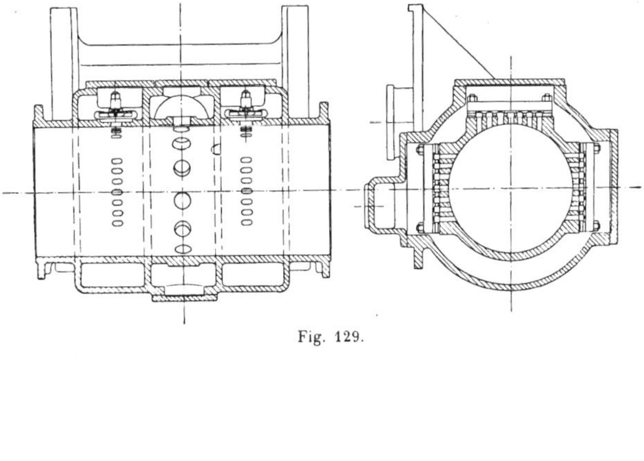

Fig. 129.

the exhaust end of the stroke. The quantity diminishes with the speed and with the pressure in the receiver. The following table gives the percentage of steam

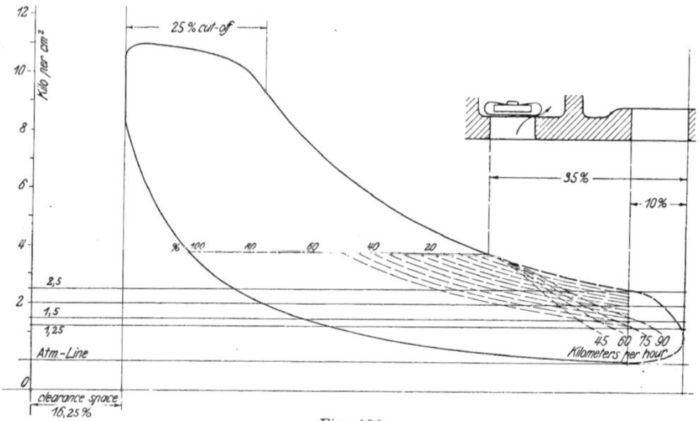

Fig. 130.

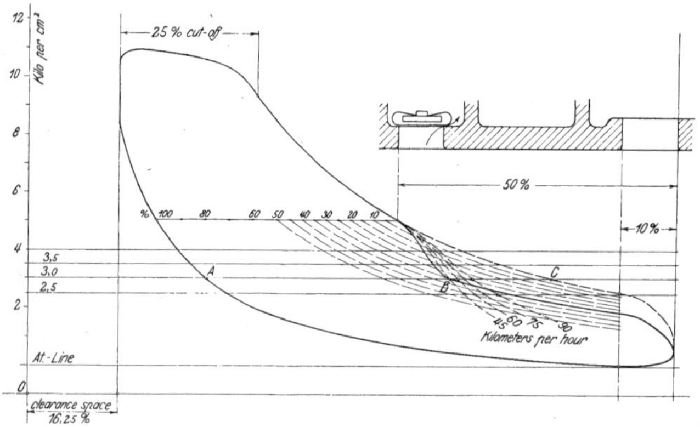

Fig. 131.

withdrawn, to the entire steam used in the cylinder, for the cases where the steam withdrawal valves are located at 35% and 50% before the end of the stroke. The variations with different speeds and steam pressures are also given.

Steam quantities withdrawn:

Pressure in receiver kgs. per cm²	Cut-off at quarter of the stroke			
	45 km per hour	60 km per hour	75 km per hour	90 km per hour
Steam withdrawal valve 35 % before the end of the stroke:				
1·25	59·0 %	55·0 %	52·0 %	43·0 %
1·5	51·0	48·5	46·0	41·0
2·0	37·5	35·2	32·5	30·0
2·5	25·0	22·5	20·0	17·0
Withdrawal valves midway in the stroke:				
2·5	41·8	39·4	37·0	33·6
3·0	29·4	27·0	25·1	22·8
3·5	20·1	18·7	17·0	15·3
4·0	11·3	10·3	9·2	7·9

On the German Railways the pressure of the heating steam has gradually been increased to 4 atmos., on account of the small cross section of the heating pipes and the ever increasing length of the trains. If the cross section of the heating pipes were doubled, whilst still giving the same normal heating effect, the pressure of steam for heating purposes would be so reduced that more than enough steam could easily be obtained for heating carriages, as a glance at the above figures will show. This would solve the problem of utilising steam which has done some work in the engine cylinder for heating purposes, whereas at present the heating of the carriages is usually effected with fresh boiler steam. The withdrawal of heating steam in this way is possible, in the case of the una-flow locomotive, because the steam consumption is small and the blast is more than enough due to the forcible exhaust.

Figure 132 (Plate 2) shows the sectional view of a 3/4 coupled una-flow locomotive. Figure 133 shows a picture of same. Two such locomotives were built by the Kolomnaer Maschinenbau-A.G. in Kolomna, for the Russian State Railway. These locomotives were arranged to work with saturated steam at the suggestion of the Author. Two sister una-flow locomotives built by the same firm are provided with superheaters. The saturated steam locomotives work with aa pressure of 14 atmos., and the superheated steam locomotives work with a pressure of 12 atmos. The increase in the case of the saturated steam engine was rendered possible by utilising the weight obtained by dispensing with the superheater for the purpose of increasing the thickness of the boiler shell, whilst the load on the axles remains the same. As, moreover, one atmosphere pressure is lost in the superheater the real difference in pressure at the engine is three atmospheres to the

Table II.

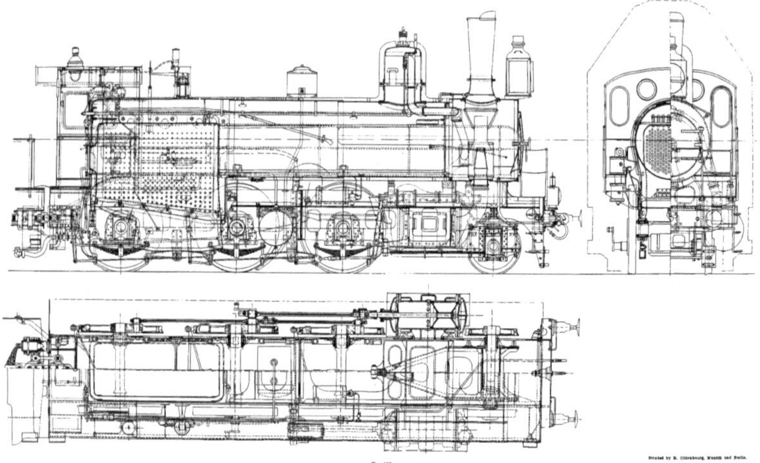

Fig. 133.

Printed by R. Oldenbourg, Munich and Berlin.

Fig. 133.

good for the saturated steam engine. The saturated steam locomotive is simpler and cheaper. All the information obtained from previous experience and tests has been employed in the design of these engines. Special attention may be directed to the careful jacketting of the cover and to the silencers. The water of condensation from the jacket is returned to the boiler by means of a small pump operated from the valve rod. The principal dimensions of the saturated steam locomotive are as follows:

Cylinder diameter	500 mm.
Piston stroke	650 mm.
Driving wheel diameter	1700 mm.
Bogie wheel diameter	1350 mm.
Steam pressure	14 atmos.
Heating surface of the boiler	166·57 sq. m.
Grate area	2·45 sq. m.

A una-flow goods locomotive for working with superheated steam is shown in figure 134. This locomotive was built for the North Railway in France. In this case it was necessary to retain the entire construction of the locomotive and gear and to replace the old counterflow cylinder by a una-flow cylinder. This was rendered more difficult by the inclined arrangement of the Stephenson link gear, but the difficulty was overcome by introducing a horizontal valve rod operated from the inclined rod through a short link as indicated in the drawing. In this engine, a silencing box was employed, into which the bends leading from the exhaust belts of both cylinders open, and from which the blast pipe ascended to the blast nozzle. A resilient valve of the form shown in figure 119 was employed in this engine. The principal dimensions of this engine are as follows:

Cylinder diameter	570 mm.
Piston Stroke	650 mm.
Driving wheel diameter	1300 mm.
Bogie wheel diameter	1040 mm.
Steam pressure	12 atmos.
Heating surface of the boiler	101·75 sq. m.
Heating surface of the superheater	30·75 sq. m.
Grate area	2·14 sq. m.
Weight (empty)	51 500 kg.
Weight (in service)	56 900 kg.
Effective weight on rails	46 900 kg.

In view of the favourable results obtained with their first una-flow locomotive, the authorities of the Moscow Kazan Railway decided to introduce five more. These additional locomotives were 5/5 coupled una-flow locomotives working with superheated steam.

Figure 135 (Plate 3) shows the principal sections of these locomotives which are specially interesting on account of their general construction and size.

In this construction, as in those described above, the fundamental principles set out by Mr. Noltein of the Moscow Kazan Railway, have been very care-

Table III.

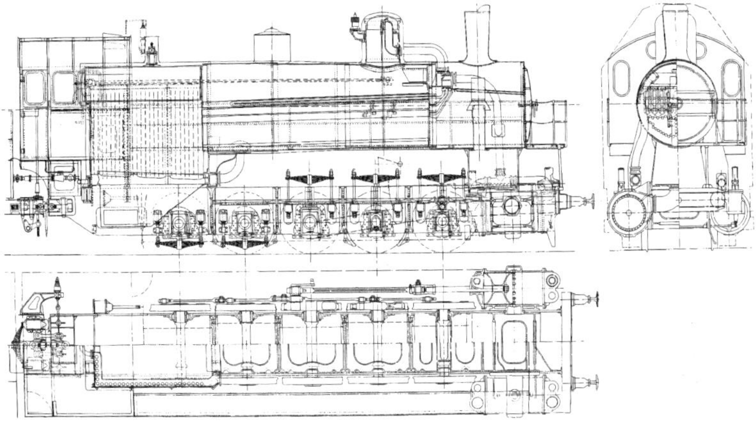

Fig. 135.

Printed by R. Oldenbourg, Munich and Berlin.

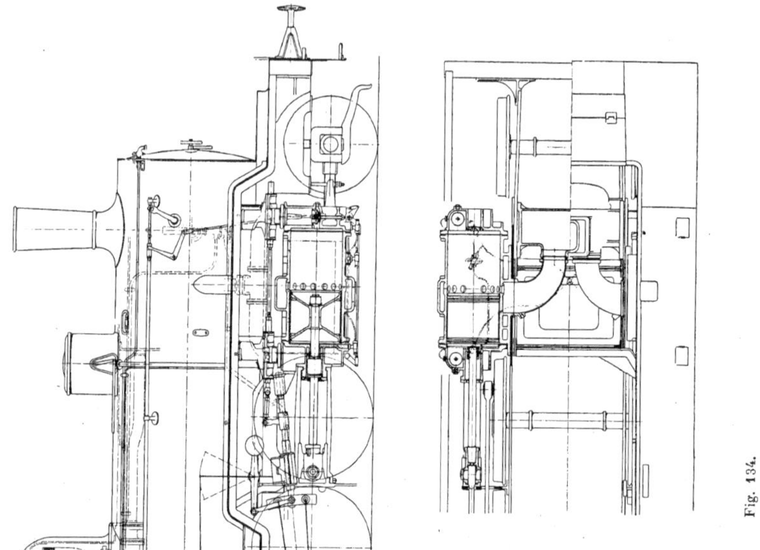

Fig. 134.

fully attended to. These principles are briefly that the vertical forces are not so detrimental to the rails as the horizontal forces. The axle weight in these locomotives is greater than the weights customary on the German railways although the rails themselves are lighter.

Mr. Noltein, the superintendent of the Moscow-Kazan Railway has employed all conceivable and clever artifices in order to reduce the unbalanced horizontal forces to a minimum. The locomotives were provided with all the latest novelties

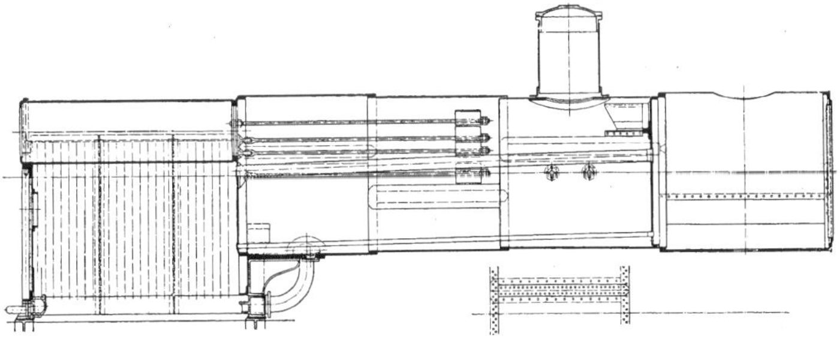

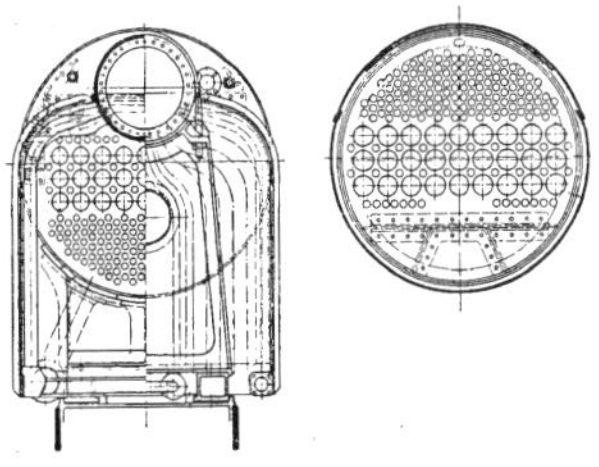

Fig. 136.

such as Schmidt Superheaters, the Goelsdorf axle arrangement, the Brotan boiler and the una-flow engine. The Brotan boiler is shown separately in figure 136.

Although previous experience with this boiler was not specially in its favour, Mr. Noltein, in his endeavour to produce a really good locomotive boiler, has left no stone unturned and works on what the author considers correct lines in maintaining that the fundamental principles of the design of this boiler are excellent. The tractive power of this locomotive is so great that some of the waggons are coupled direct to the locomotive by a wire rope, instead of being coupled to the immediately preceding waggons. Fifty seven waggons are following those which are coupled to the locomotive by means of the wire rope and this figure is considered quite permissible for the couplings according to the strength tests carried out on the waggons in the Moscow-Kazan Railway.

The principal dimensions of these locomotives are as follows:

Cylinder diameter	680 mm.
Piston Stroke	700 mm.
Driving wheel diameter	1320 mm.
Steam pressure	12 atmos.

Table IV.

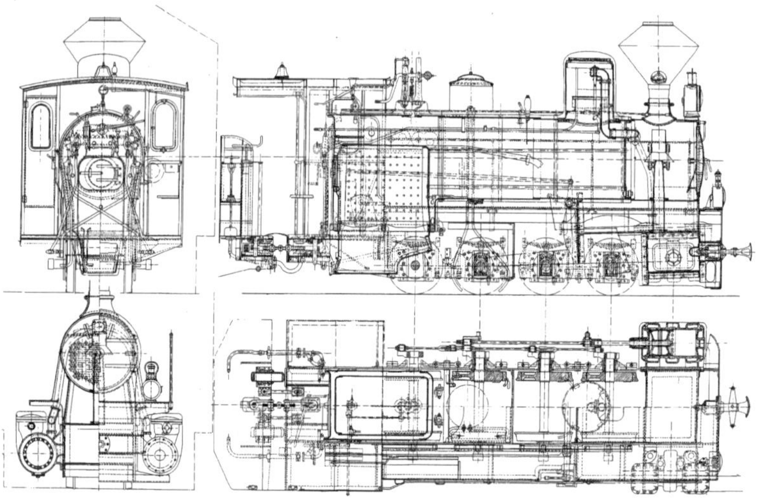

Fig. 137.

Printed by R. Oldenbourg, Munich and Berlin.

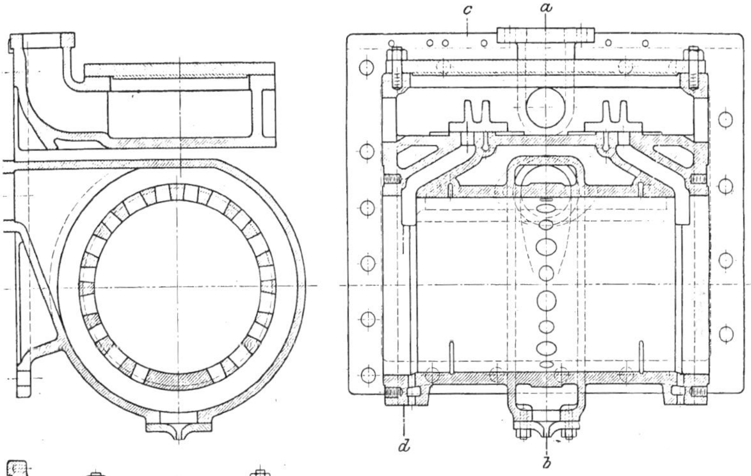

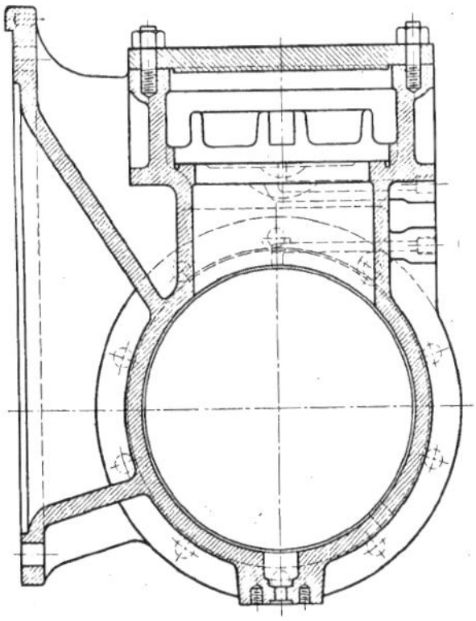

Fig. 138.

Heating surface of the boiler	211·52 sq. m.
Heating surface of the superheater.	60·58.
Grate area	3·68 sq. m.
Weight (empty)	72 500 kg.
Weight (in service)	81 100 kg.
Effective weight on rails . .	81 100 kg.

A narrow gauge una-flow locomotive built by the Maschinenbau A.-G. Kolomna for the Turin Exhibition is shown in figure 137 (Plate 4). The una-flow cylinders of this engine are provided with slide valves in view of the small dimensions. The construction of the cylinder is separately illustrated in figure 138. As there is no exhaust recess on the slide valve, the valve pressure and the load on the gear is considerably diminished.

The lubricating oil is pressed between the rubbing surfaces so that everything is done in order to remove all possible defects consequent upon the use of a flat valve. The usually large power which the valve gear of a locomotive is called upon to exert and the consequent wear and tear on the valve gear is considerably diminished with this construction. The engine is designed to work with saturated steam and in view of this, both the cover and the cylindrical walls near the inlet ends, are steam heated.

The principal dimensions of this locomotive are as follows:

Cylinder diameter	355 mm.
Piston Stroke	350 mm.
Driving wheel diameter	750 mm.
Steam pressure	12 atmos.
Heating surface of the boiler	54·26 sq. m.
Grate area	0·93 sq. m.
Weight (empty)	19 200 kg.
Weight (in service)	21 200 kg.

A four cylinder una-flow express locomotive engine is shown in figure 139. In this case two adjacent cylinders with cranks at 180⁰ are operated from a single admission valve gear arranged in the ordinary manner. The two adjacent cylinders have two cross pipes leading to opposite ends of the cylinders and retaining these ends of the cylinders always in communication. These cross pipes are so dimensioned that they provide the necessary clearance space. The clearance is accomodated in the cross connections and not in the cylinder. Each cylinder is provided with a ring of exhaust ports controlled by the piston and the exhaust belts are all joined by the frame. In the centre of the cylinder casting a single ascension pipe is provided which carries the exhaust steam to the blast. This construction has the great advantage of simplicity. As the two inner cylinders have no special valve gear, there are only two gears such as are used on any two cylinder locomotive. As all the cylinders are of the same diameter and stroke, the inertia diagrams are smaller, and the balancing is perfect, with the exception of the unbalanced forces due to the finite connecting rod. The steam diagrams are all similar, so that the turning efforts are always equal and uniform.

The dimensions of these locomotives are as follows:

Cylinder diameter	430 mm.
Piston stroke	600 mm.
Driving wheel diameter	2100 mm.
Bogie wheel diameter	1000 mm.
Steam pressure	12 atmos.
Heating surface of the boiler	126·98 sq. m.
Heating surface of the superheater . . .	40·32 sq. m.
Grate area	2·31 sq. m.
Weight (empty)	62 900 kg.
Weight (in service)	65 000 kg.
Weight effective on rails	35 000 kg.

Another una-flow locomotive cylinder is shown in figures 140 and 141. In this design the steam distribution valves are of the piston type and are arranged in two parts located near each end of the cylinder. The valve chest in this case extends over the entire length of the cylinder which gives a kind of "air vessel"

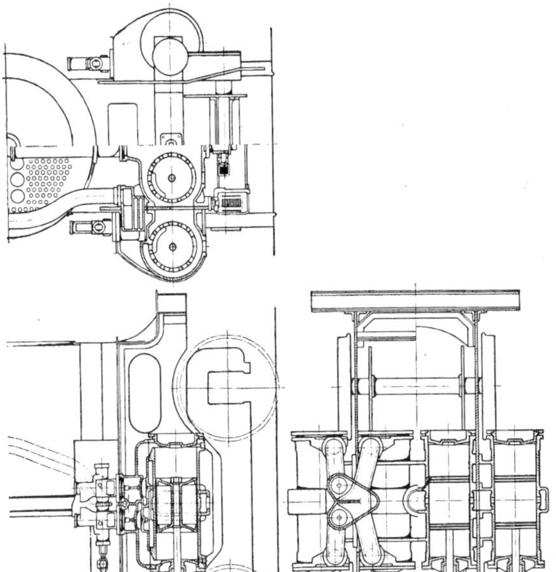

Fig. 139.

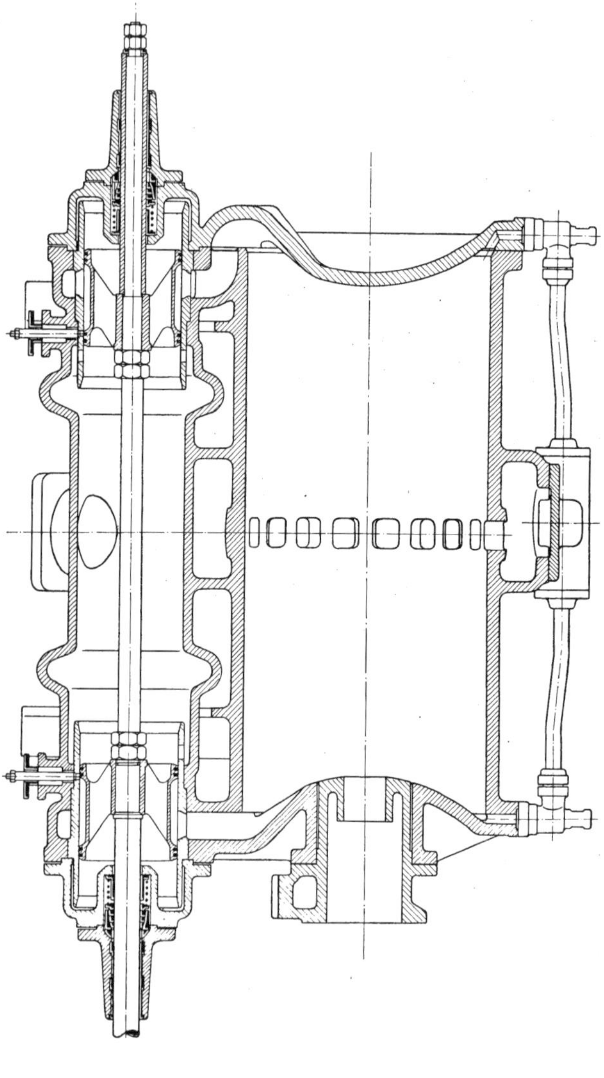

Fig. 140.

action owing to the large steam space located between the halves of the piston valve. It will be noticed that the double seated relief valve described earlier is also provided in this design and is located at the lower part of the cylinder. The relief valve is controlled from the link so that when the link is set to give late cut-offs, such as at 50% to 75% of the stroke, the relief valve becomes operative to reduce the compression. This is effected by coupling the valves in the branch

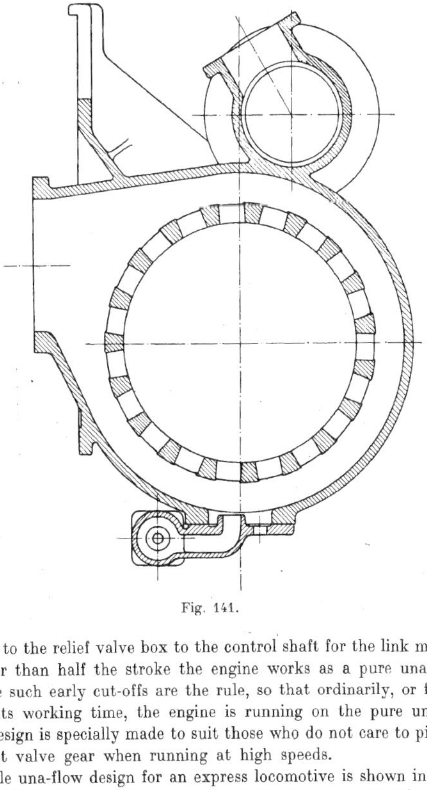

Fig. 141.

pipes leading to the relief valve box to the control shaft for the link motion. With cut off earlier than half the stroke the engine works as a pure una-flow engine and of course such early cut-offs are the rule, so that ordinarily, or for the great majority of its working time, the engine is running on the pure una-flow principle. This design is specially made to suit those who do not care to pin their faith to the poppet valve gear when running at high speeds.

A suitable una-flow design for an express locomotive is shown in Figures 142 and 143. In this design special measures are taken to reduce the clearance to the usual amount. In a pure una-flow engine working with superheated steam at

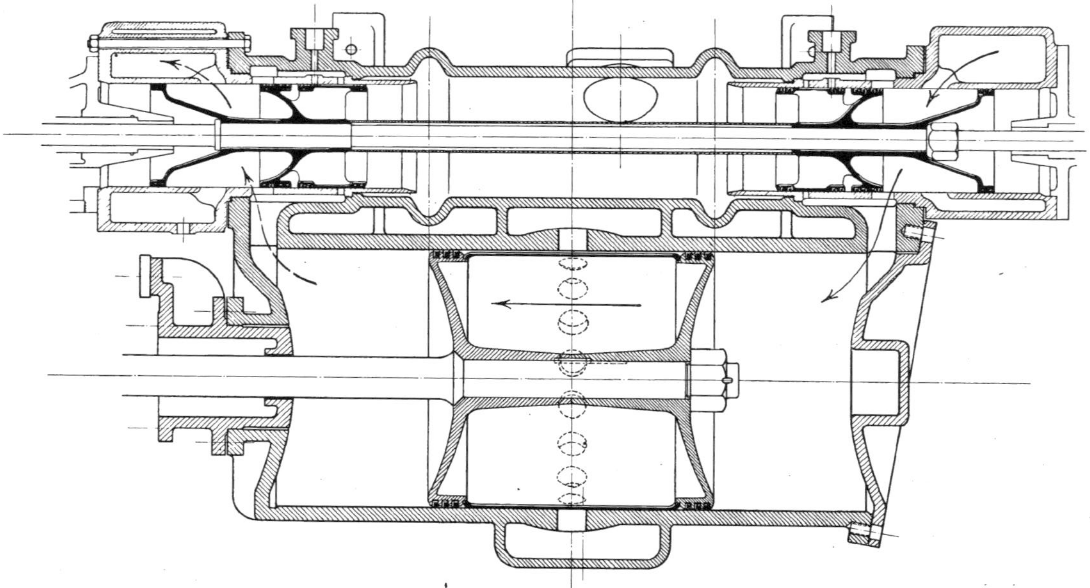

Fig. 142.

12 atmos. pressure, the clearance space should be about 16%, to prevent the compression pressure rising above the admission pressure. The design shown in figures 142 and 143 is intended to provide a supplementary clearance amounting to about 25% of the volume swept out by the piston. This supplementary clearance is accomodated in the cover castings of the piston valve casing. At

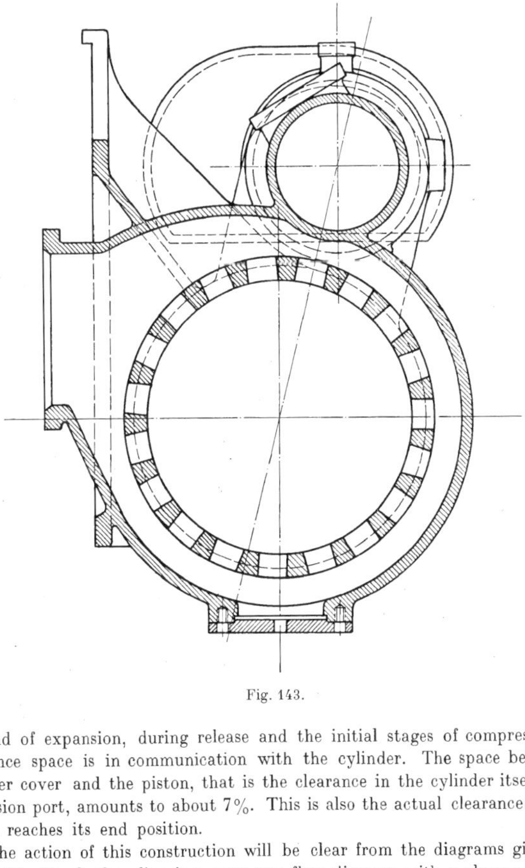

Fig. 143.

the end of expansion, during release and the initial stages of compression, this clearance space is in communication with the cylinder. The space between the cylinder cover and the piston, that is the clearance in the cylinder itself and the admission port, amounts to about 7%. This is also the actual clearance when the piston reaches its end position.

The action of this construction will be clear from the diagrams given in figure 144. The broken line is a pure una-flow diagram with a clearance of 16%

and cut-off at $^1/_5$ th. There is also shown in full lines the diagram for a una-flow engine working with a supplementary clearance amounting to 25% and cut-off at $^1/_5$ of the stroke.

The piston valve has a normal admission lap but a negative exhaust lap amounting in this case to 20 mm. The supplementary clearance space is in communication with the interior of the cylinder simultaneously with the uncovering of the exhaust ports by the piston. The early compression is into a space composed of the cylinder clearance proper (7%), and the supplementary clearance (25%). In the diagram the compression line is plotted from the point O_1. The compression

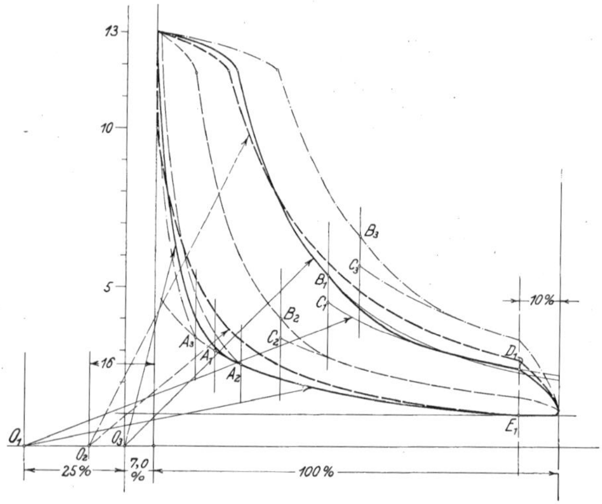

Fig. 144.

under the conditions mentioned lasts up to the point A_1, where the supplementary clearance space is cut out by the motion of the valve and the compression continues further into the 7% clearance of the cylinder and ports. This later part of the compression line is plotted from the point O_3. The una-flow compression line with a 16% clearance is plotted from the point O_2. After compression is completed the admission takes place — which in both cases continues up till 20% of the stroke has been completed. The initial pressure is taken at 13 atmos. absolute. The initial part of the expansion line is drawn from the point O_3. This expansion continues under these conditions up to the point B_1, where communication is established between the supplementary clearance and the interior of the cylinder. At this point the volume of the supplementary clearance space is added to the volume of the cylinder and there is a mixture with the compressed steam left in the supplementary clearance at the point A_1. The pressure of this mixture is indicated by the point C_1. This pressure, however, is not attained as the high velocity of the piston and the throttling due to the gradual opening of the connecting passage from the supplementary clearance space to the cylinder cause a gradual fall. The actual expansion line gradually runs into the common expansion line of the mixture drawn from the point C_1. From this it will be observed that the actual expansion line soon coincides with the expancion line drawn from C_1. This expansion continues until the point D_1, when the exhaust ports are opened and steam escapes from the cylinder and supplementary clearance space. From the point E_1, the cycle of operations described is repeated. With cut-off at 10% of the stroke, the point A_1, is displaced to the right in the diagram to the point marked A_2. The point C_1, with cut-off at 10% of the stroke is displaced to the point C_2. With cut-off at 30% of the stroke the points A_1 and C_1 are displaced to the points A_3 and C_3 respectively.

The variation in the compression is not so great as in ordinary locomotives, but is nevertheless present to an extent, such as is necessary to meet the requirements of good steam distribution. This will be dealt with more fully in the next chapter.

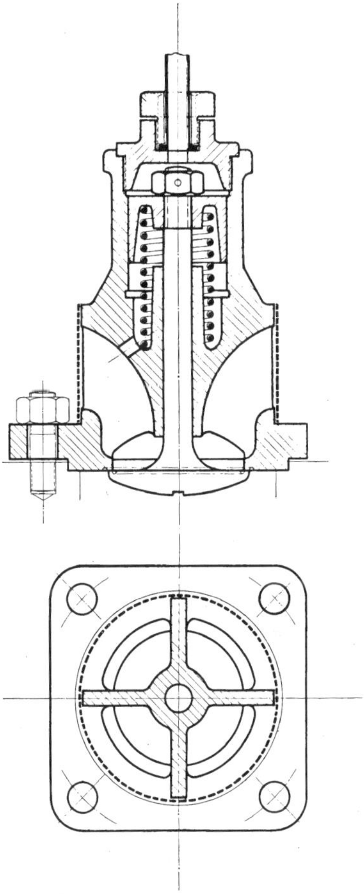

Fig. 145.

In ordinary locomotives the last part of the expansion line falls away owing to an excessively early release. As can be seen from the diagram in figure 144 the new method of steam distribution avoids this by the addition of the compressed steam in the supplementary clearance space. A considerable area is also saved on the compression side. The expansive energy of the steam, as will be noted from the position of the point D_1, is more effectively utilised due to the use of the small actual cylinder clearance space of 7%.

Another feature of this construction is that the leakage losses are diminished. The central space between the two piston valves is filled with boiler steam. At each end of this central space there is arranged a compression chamber which is connected through ports to an annular compression chamber or clearance in the cover casting of the valve chamber. The compression pressure in the end chambers on the piston valve amounts to 2 to 3 atmos. Beyond this compression chamber there is a further chamber at each end of the valve chest, and this extreme outer chamber may advantageously be placed in communication with the exhaust belt around the central part of the cylinder. There will then be atmospheric pressure. The packing gland for the valve spindle may therefore be dispensed with, and the spindle may be led through a simple neck bush. The above construction gives what might be called a compound packing for the exhaust side of the valve, and this exhaust side is the most important part of the valve to pack. Any steam leaking past the valve enters the supplementary clearance and is again used to perform useful work in the cylinder during the latter part of the expansion. There are two features of importance in this construction, the first of which is that the usual and necessary 16% clearance is dispensed with and in place thereof a small 7% clearance in the cylinder is used in combination with a supplementary clearance which is brought into communication with the cylinder during the latter part of the expansion and the early part of compression. The second feature of importance is that the leakage losses of the valve are diminished by the compound or multistage arrangement of the packing. The entire engine might be said to work in a kind of compound manner. Experiments have shown that the principal advantage is obtained by the improvement in the packing. It is also of some importance to note that the uni-directional flow principle is maintained throughout the whole action of the engine. The steam in the additional clearance space oscillates backwards and forwards, but there is no real withdrawal such as takes place in the ordinary counter-flow engine. There is little or no cooling or extraction of heat from the clearance surfaces because it is always practically the same quantity of steam which pendulates backwards and forwards between the compression space and the interior of the cylinder.

The construction just described has possibly its greatest sphere of usefulness in locomotives working with a moderate boiler pressure. Such pressures would necessitate a considerable clearance space in the cylinder and by the arrangement just described the cylinder clearance could be reduced whilst still retaining a pure uni-directional flow action. Two express engines on the Prussian State Railways have been built to this design.

In these engines air suction valves were employed, such as shown in Figure 145. These valves are designed by the Knorrbremsen A. G. in Berlin-Boxhagen. The purpose of these valves is to admit fresh air to the cylinder during coursing, so as to avoid excessive heating of the interior of the cylinder due to the back and forward pendulating movement of the steam from end to end of the cylinder during coursing. This back and forward movement of the steam might cause carbonisation of the lubricating oil. Compressed air or steam is admitted through the uppermost connection from a valve controlled from the cab. By introducing compressed air or steam the piston in the upper chamber is pressed downwardly against the pressure of a spring and thereby the air suction valve is retained open. Around the valve casing there is arranged a sieve to exclude all dust or foreign matter. With a valve of this kind arranged on the cylinder, air from the atmosphere is drawn into the cylinder at each stroke when the engine is coursing and is again expelled from the cylinder to the atmosphere. This retains the interior of

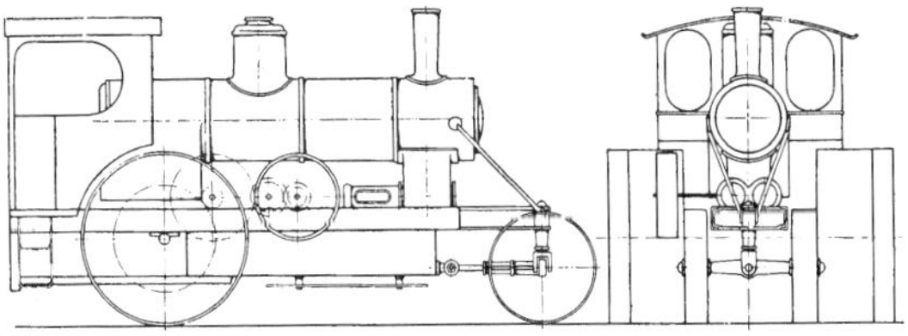

Fig. 146.

the cylinder at a normal temperature and prevents all carbonisation of the oil. The valve may be arranged on the steam supply pipe as near as possible to the cylinder.

In figure 146 there is shown a una-flow traction engine. This engine is designed for drawing waggons, ploughs or any other agricultural machines.

In the case of this traction engine, no superheating was employed so as to simplify the construction and in view of the advantages obtained by employing the una-flow cylinder.

The principal dimensions of this traction engine are as follows:

Cylinder diameter	255 mm.
Piston stroke	280 mm.
Revs. per minute	255
Driving wheel diameter	2000 mm.
Bogie wheel diameter	1300 mm.
Steam pressure	10 atmos.
Heating surface of the boiler	25·3 sq. m.
Grate area	0·72 sq. m.
Weight (empty)	13 000 kg.
Weight (in service)	16 500 kg.

Chapter IX.

The influence of the clearance volume on the steam consumption.

The shaded parts of the diagram shown in figure 147, represent the losses due to the clearances in the cylinder and ports. In drawing the expansion and compression lines, the law of Mariotte has been followed. What the diagram

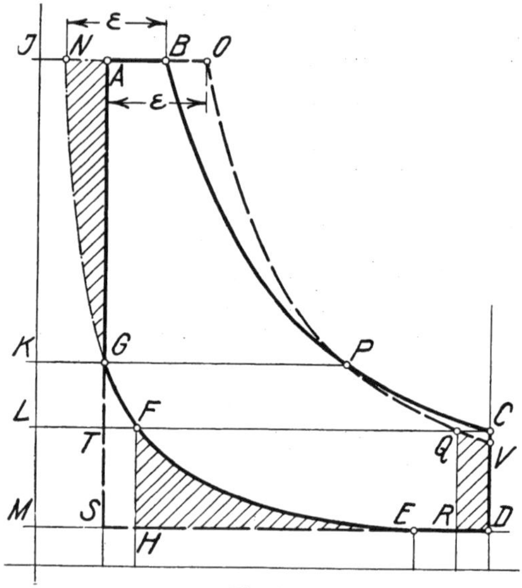

Fig. 147.

really shows is the power obtained with a clearance of nil as compared with the power obtained with a clearance amounting to 17% of the volume swept out by the piston at each stroke. The same quantity of steam exactly is assumed to form the working medium in each case. The length *NB* is accordingly equal to the length *AO* — it being assumed that the pressure at *N* is equal to the theoretical end pressure of compression. The full line diagram *ABCDEG* is for the 17% clearance and the dotted line diagram *AOVDS* is for the case where the clearance is nil. The losses due to the clearance are thus represented by the three shaded parts shown on the figure. The area *NBPG* = area *AOPG*. The area *NAG* lies outside the piston stroke and may therefore be taken as quite lost. The area

GPCF = area *GPQT* so that there is here no loss. The area *FCDH* = area *TQRS* which shows a loss represented by the two shaded parts *FEH* and *QVDR*. With later cut-offs the area of the part representing the loss at the exhaust end may fall to nil and the vertical line at the left of the diagram also forms one side of the lost area *FEH* (see the line *GS* in figure 147a). If the cut-off is made still later a new lost area creeps in (see figure 147b), but at the same time the lost area at the end of exhaust is reduced.

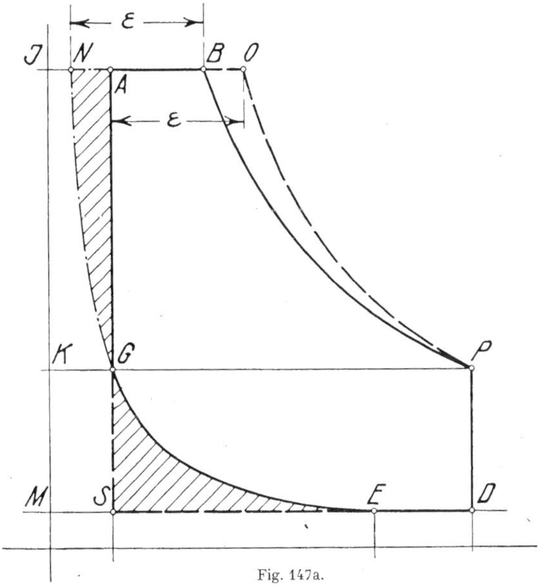

Fig. 147a.

The comparison of the diagrams shows that with a clearance of 17% the clearance losses are considerable but with a clearance equal to nil the clearance losses are also equal to nil.

The method of reasoning would be the same if we had taken adiabatic instead of the Mariotte lines for expansion and compression. No great error occurs in using the Mariotte lines for counter-flow engines working with saturated steam, although in such engines the compression line shows a tendency towards the adiabatic.

The following important conclusions may be safely drawn from a consideration of figures 147, 147a, 147b.

1. The amount of the clearance losses depends on the volume of the clearance.
2. The amount of the clearance losses depends upon the amount of steam admitted to the cylinder.
3. The amount of the clearance losses depends upon the beginning of compression.
4. The clearance losses will be a minimum when the total area of the shaded parts is a minimum. For a given clearance and a given amount of steam

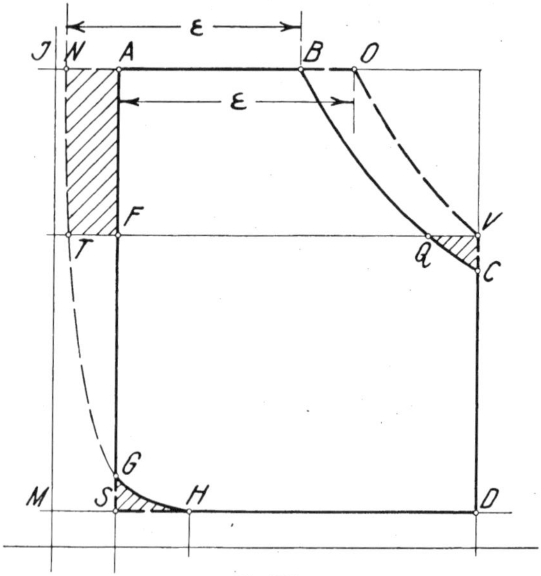

Fig. 147 b.

admitted to the cylinder, this minimum may be easily secured by properly timing the compression.
5. In all cases the pressure at the end of compression must not exceed the initial pressure.

The rules which the designer should adopt may be summarised as follows. *The volume of the clearance should be kept as small as possible and this volume should be accomodated with a minimum amount of surface and this minimum amount of surface should be treated thermally as favourably as possible by using the uni-directional flow. Then the compression should be timed so that the clearance losses are a minimum.*

The view commonly held that the clearance losses may be eliminated by increasing compression so that the clearance is filled with compressed steam at the admission pressure, is only true if the drop of pressure during exhaust is equal to nil. But even in this case a loss occurs as the effective volume of the cylinder is increased from v_1 to v_2 (see fig. 147c).

The amount of the clearance losses in relation to the cut-off may be determined in the following manner for any given size of clearance.

In figure 148, the effective diagram area $F = ABEDCF$ may be taken as composed of

$$ABHG + BEKH - FCJG - CDKJ;$$

call this

$$f_1 + f_2 - f_3 - f_4 = F \quad . \quad . \quad . \quad . \quad . \quad . \quad . \quad . \quad . \quad . \quad (1)$$

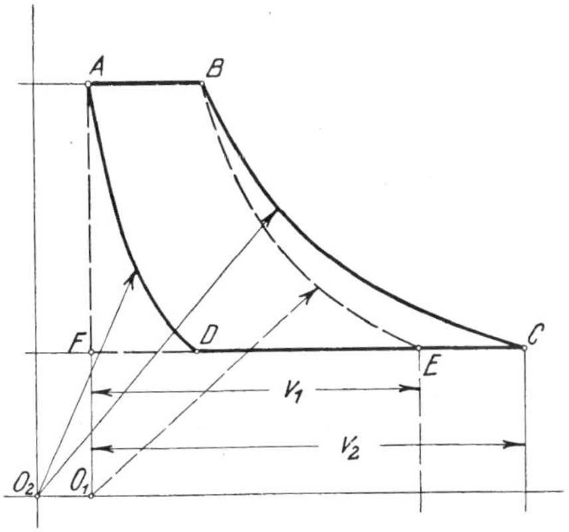

Fig. 147 c.

The quantity of boiler steam supplied to the cylinder at each stroke is represented by

$$k = S_o + V_e' - V'_c \text{ (in figure 148)} \quad . \quad . \quad . \quad . \quad . \quad . \quad . \quad (2)$$

This quantity will, for the purposes of this investigation, be assumed as constant. The amount V_c' depends upon the length of the compression represented by V_c at the foot of the diagram. Thus

$$V_c' p_1 = (S_o + V_c)\, p_2$$

Equation 2, above may therefore be written down in the form

$$k = S_o + V_e - (S_o + V_c)\,\frac{p_2}{p_1}$$

or

$$S_o + V_c = (S_o + V_e - k)\,\frac{p_1}{p_2} \quad . \quad . \quad . \quad . \quad . \quad . \quad . \quad . \quad (3)$$

If the length of the compression V_c is altered, the length k is displaced and the cut-off or amount of filling V_e of the cylinder is also altered. It is evident that for the maximum of economy

$$\frac{dF}{dV_e} = 0.$$

In figure 148

$$f_1 = p_1 V_e;$$

$$f_2 = \int_{(S_o + V_e)}^{(S_o + V^h)} p\,dv = (S_o + V_e)\,p_1 \log_e \frac{S_o + V_h}{S_o + V_e};$$

$$f_3 = \int_{S_o}^{(S_o + V_c)} p\,dv = (S_o + V_c)\,p_2 \log_e \frac{S_o + V_c}{S_o};$$

$$f_4 = (S_o + V_h)\cdot p_2 - (S_o + V_c)\,p_2.$$

Taking equation 3 above

$$f_3 = (S_o + V_e - k)\frac{p_1}{p_2}\cdot p_2 \log_e \frac{(S_o + V_e - k)\frac{p_1}{p_2}}{S_o};$$

$$f_4 = (S_o + V_h)\,p_2 - (S_o + V_e - k)\frac{p_1}{p_2}\,p_2.$$

Inserting these values in equation (1) we obtain

$$F = p_1 V_e + (S_o + V_e)\,p_1 \log_e \frac{S_o + V_h}{S_o + V_e} - (S_o + V_e - k)\,p_1$$

$$\times \log_e\left(\frac{S_o + V_e - k}{S_o}\,\frac{p_1}{p_2}\right) - (S_o + V_h)\,p_2 + (S_o + V_e - k)\,p_1 \quad . \quad . \quad (4)$$

Differentiating for V_e we obtain on simplifying out

$$\frac{dF}{dV_e} = p_1 \log_e \frac{S_o + V_h}{S_o + V_e} - p_1 \log_e\left(\frac{S_o + V_e - k}{S_o}\cdot\frac{p_1}{p_2}\right) = 0$$

or

$$\frac{S_o + V_h}{S_o + V_e} = \frac{S_o + V_e - k}{S_o}\cdot\frac{p_1}{p_2} \quad . \quad . \quad . \quad . \quad . \quad . \quad . \quad . \quad . \quad (5)$$

Substituting the values found in equation 3, we get finally

$$S_o + V_c = S_o \frac{S_o + V_h}{S_o + V_e} \quad . \quad . \quad . \quad . \quad . \quad . \quad . \quad . \quad . \quad . \quad (6)$$

The final equation (6), above is the expression for a symmetrical or erect hyperbola and may be easily represented graphically (fig. 148). Join the point D to the point B produce DB to cut the vertical in L; join the point L to point A and produce LA to cut the horizontal from D in R. The point R indicates the point at which the most favourable compression should begin with cut-off at the point B. From equation 6, it also follows that with steam admission up to the end of the stroke the compression should be nil and with a steam admission up to the beginning of the stroke the compression should be 100%.

Compression here is meant to indicate the portion of the stroke from the commencement to the end of compression.

From this it follows that the constant compression, always held to be a desideratum in large engines is fundamentally false. It further follows that the distribution obtained by link motion and shifting eccentric gear is fundamentally correct as it gives large compressions with early cut-offs and small compressions with late cut-offs.

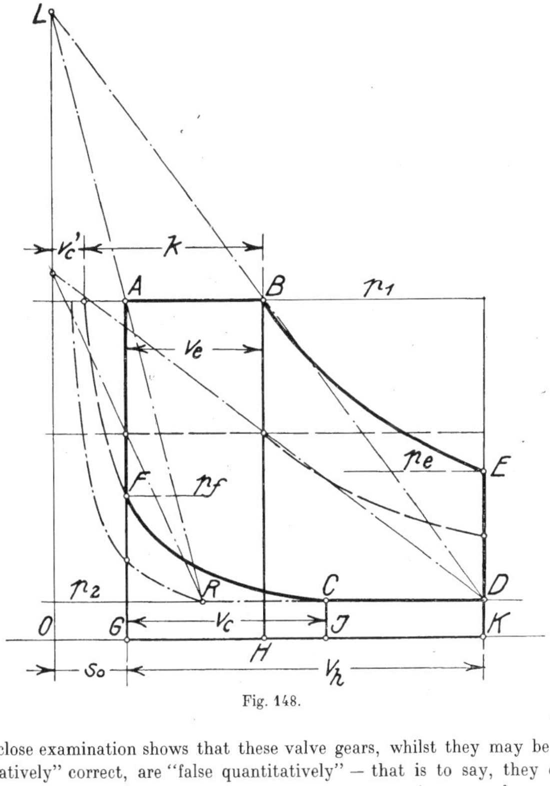

Fig. 148.

A close examination shows that these valve gears, whilst they may be called "qualitatively" correct, are "false quantitatively" — that is to say, they do not alter the compression quite in a manner to satisfy the requirements of equation 6. This will be appreciated when it is recalled to mind that a slide valve, with an exhaust lap $= 0$, gives compression for 50% of the stroke when the cut-off is at 0 %. According to equation (6), this compression should be 100%. The reason of this erroneous practice is probably to be found in the anxiety of the designer to accomodate the compressed steam in a reasonable clearance volume. The direct application of equation (6) is of course limited by the inexorable requirement that the end pressure of compression must on no account exceed the admission steam pressure.

Superheated steam 13 at. abs. 300 ° C Non-condensing, Clearance space 15,4 %

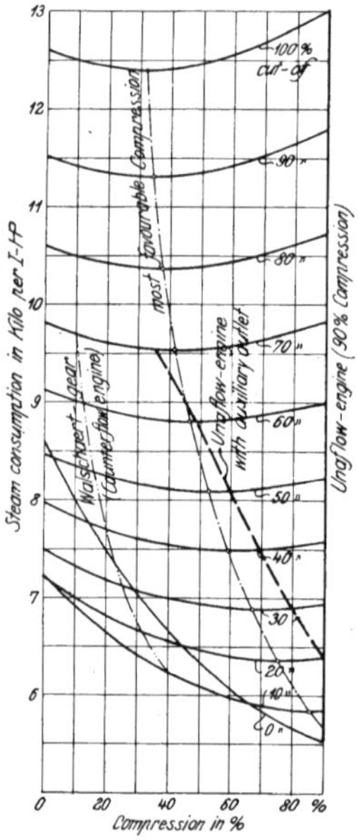

Fig. 149.

Superheated steam 16 at. abs. 300 ° C Non-condensing, Clearance space 11,9 %

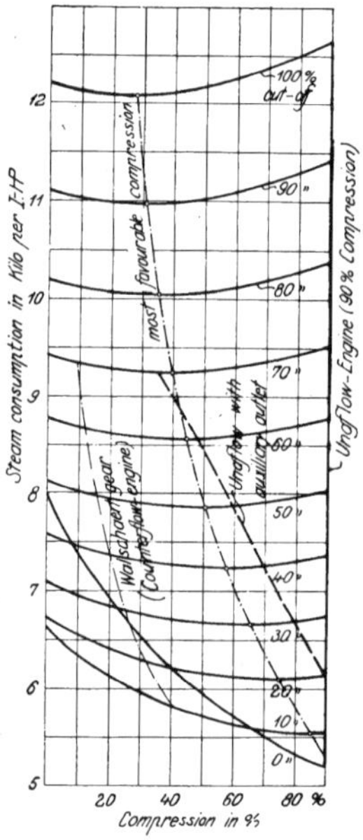

Fig. 150.

It ought also to be borne in mind that the Mariotte lines were used in the diagram. Later investigations point to the conclusion that the substitution of an adiabatic line for the Mariotte line leads to earlier commencement of compression.

In deriving the equation (6) from equation (3) it will be noted that the expression $\frac{p_1}{p_2}$, disappears, which points to the conclusion that the most favourable compression is independant of the pressure fall. With constant cut-off and constant clearance, the most favourable compression is quite independent of fluctuations in the pressure fall (fig. 148). In a compound engine with constant cut-off in the low pressure cylinder, any variation in the cut-off in the high pressure cylinder will cause an alteration in the pressure fall in the low pressure cylinder. In this case the compression in the low pressure cylinder is in no way affected because the cut-off is constant, but in the high pressure cylinder, an alteration (diminution) of the compression is necessary in view of the later cut-off.

Saturated steam 13 at. abs. Non-condensing, Clearance space 13 %

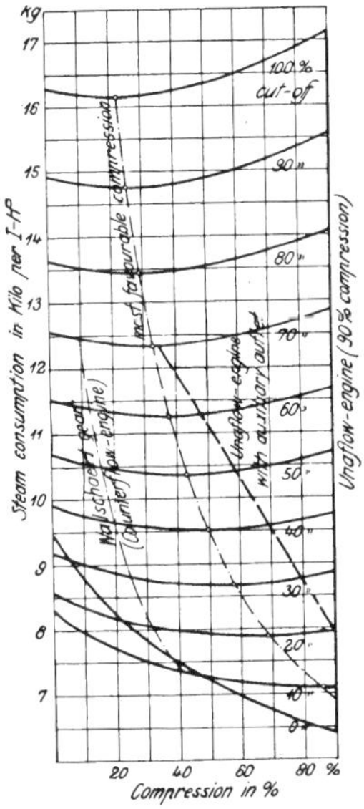

Fig. 151.

Saturated steam 16 at. abs. Non-condensing, Clearance space 10,1 %

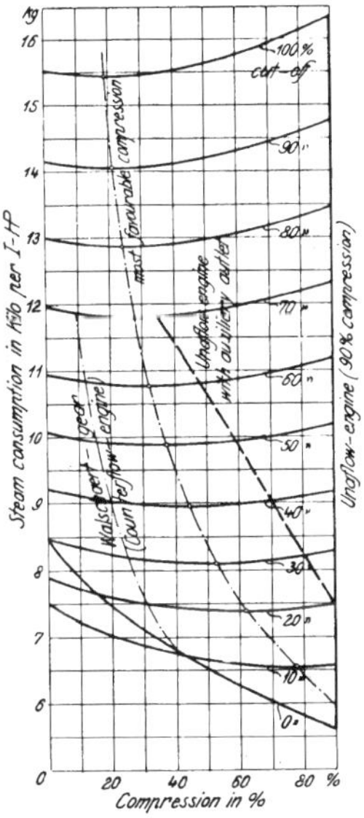

Fig. 152.

In figures 149—152, diagrams are shown representing the ideal steam consumption per I. H. P., with various compressions and cut-offs and with exhaust to atmosphere. The latest and best available entropy charts were employed in deriving these diagrams which are drawn chiefly with a view to locomotive practice.

The points of most favourable compression for all cut-offs are indicated by the broken line curves which are marked "*most favourable compression*". These curves show, in the case of figures 149—150; that with cut-off at about 5% of the stroke, the most favourable compression is 90%. As a compression extending over more than 90% of the stroke is scarcely practicable, this may also be taken as the most favourable compression for all cut-offs earlier than 5%. The curves also show, however, that the 90% compression of the una-flow engine does not involve a great increase in steam consumption, even when the cut-off rises to 30%—40% as is customary in practice. The compression values (90%) for the

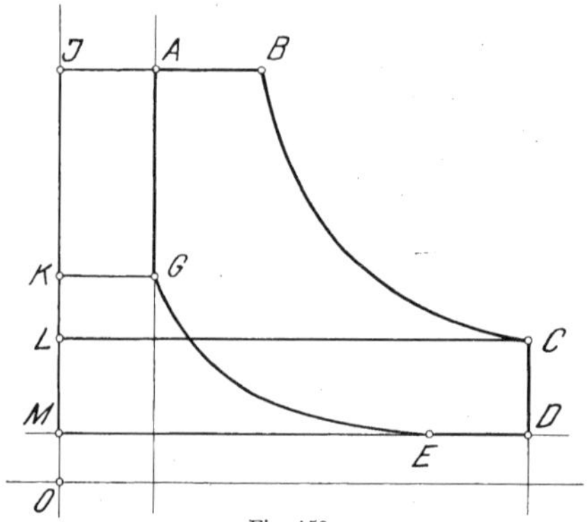

Fig. 153.

una-flow engine are given by the extreme right hand line which is marked "una-flow engine (90% compression)". For purposes of comparison, the compressions obtained by the "Walschaert" gear, as used in the counter-flow locomotives of the Prussian State Railways, are also drawn in. It wil be noticed that the compressions obtained with the Walschaert gear are fairly

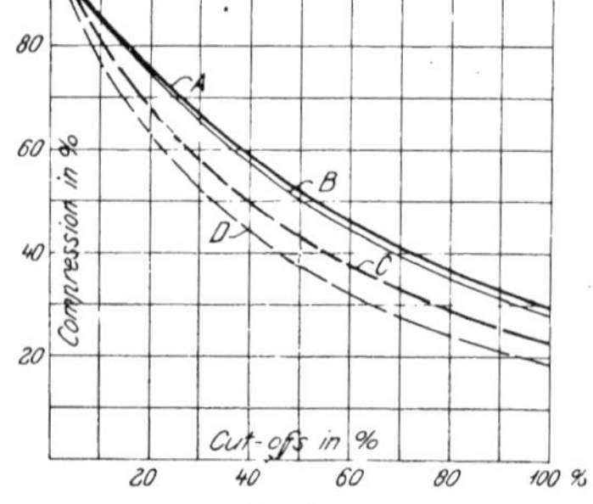

Fig. 154.

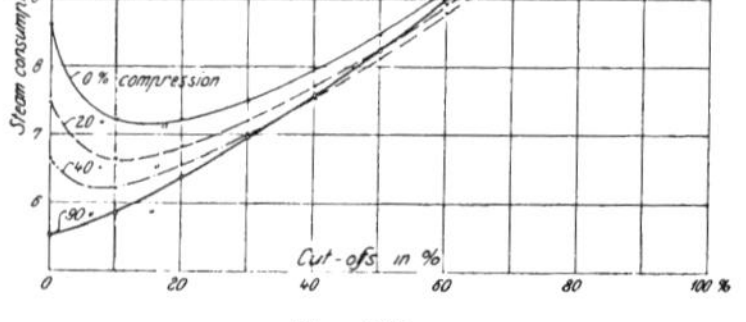

Fig. 155.

remote from the most favourable compression. Up to cut-off at 60% a una-flow engine with 90% compression has the advantage over the Walschaert gear, but from cut-off at 60% upwards the advantage lies with the "Walschaert". In figure 150, the una-flow engine has the advantage in this respect up to cut-off at 40% of the stroke and thereafter the advantage rests with

Superheated steam 16 at. abs. 300° C Non-condensing Clearance space 11,9%

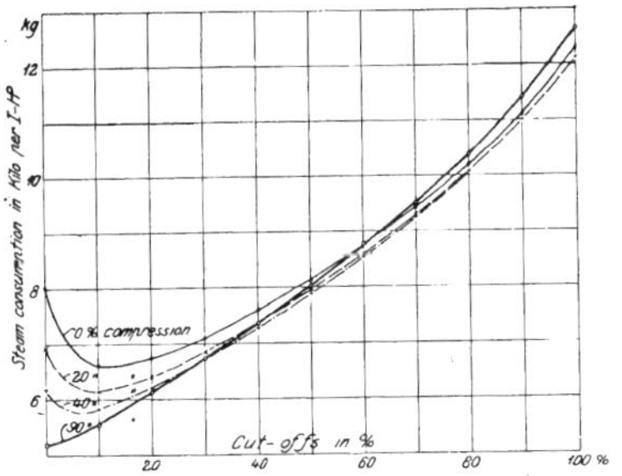

Fig. 156.

Saturated steam 13 at. abs. Non-condensing Clearance space 13%

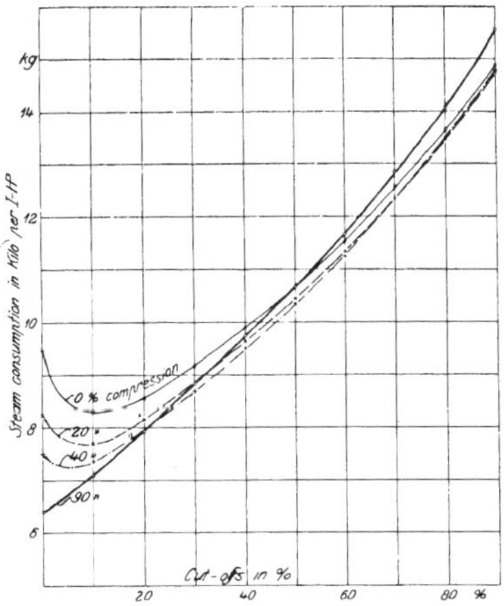

Fig. 157.

Saturated steam 16 atm. abs. Non-condensing Clearance space 10,1%

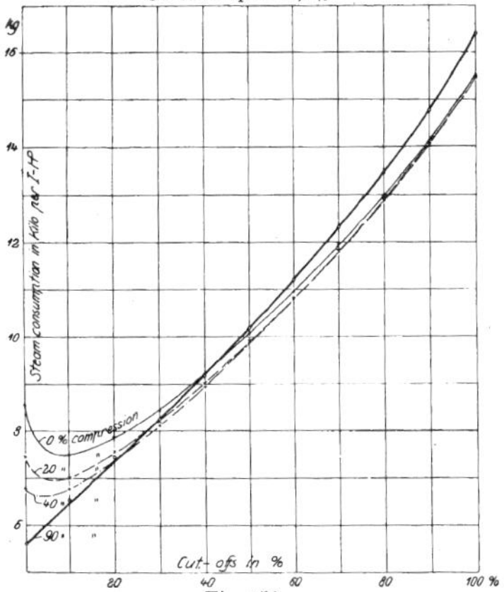

Fig. 158.

Superheated steam 13 at. abs. 300° C Non-condensing

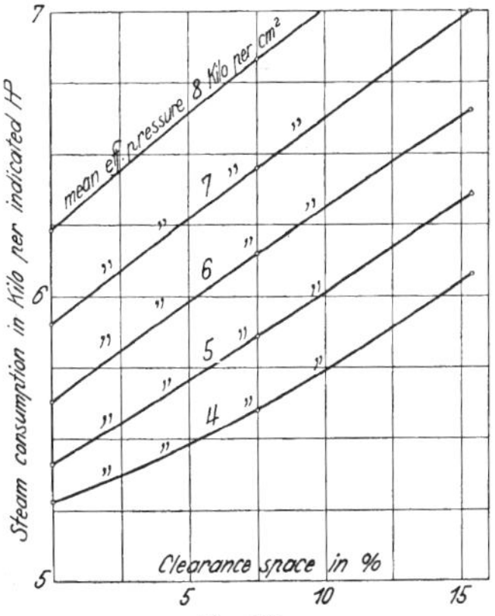

Fig. 159.

the Walschaert gear. The crossing point of the advantage as regards most favourable compression in the cases shown in figures 151—152 is with cut-off at about 30%.

As cut-offs later than 35% of the stroke only temporarily occur in locomotive practice, the above considerations show that, as regards the most favourable compression, the advantage rests to a small extent with the una-flow engine as compared with the ordinary counter-flow. This is assuming of course the same clearance in both engines.

In each of the diagrams (figures 149—152) there has also been inserted a curve representing the compressions obtained in a una-flow engine provided with an auxiliary exhaust. These curves are marked "una-flow engine with auxiliary outlet". The exhaust lap on the auxiliary valve is taken to be such that up to cut-off at 20% of the stroke the engine works with a pure una-flow action and the full una-flow compression of 90%. With cut-offs later than 20% the auxiliary valve comes into operation and reduces the compression as the cut-off gets later. It will be seen that the curve for the una-flow engine with auxiliary exhaust slopes off in a similar manner to the ideal curve, and, in figures 149-150, it comes very close to the ideal.

Another interesting point to note, and it may be seen in all the diagrams, is that the steam consumption, with cut-off at 0% (that is with the clearance space *only* filled with steam) is with *low compressions* greater than the steam consumption with cut-off at 10%, 20%, or 30% and in exceptional cases is even in excess of that with cut-off at 40% or 50%. This is in conformity with equation 6.

The steam consumption figures in diagrams 149 to 152 have been derived from the entropy temperature chart on the basis of the diagram shown in figure 153. In this figure it is assumed that

$$JBCL + LCDM - JAGK - KGEM = ABCDEG.$$

this last area is a measure of the ideal steam consumption. The values for the factors in the above equation may be taken directly from the entropy chart.

The results obtained from the diagrams shown in figures 149 to 152 have been summarised in figure 154. In this figure, the values of the most favourable compressions in the four cases mentioned are drawn with reference to various cut-offs. From the curves *ABCD* it will be seen that with superheated steam a higher compression is necessary for any given cut-off than with saturated steam, other things being equal. It follows from this that the una-flow locomotive must approach more closely to the most favourable compression for superheated steam, than for saturated steam. Another point to be noted with regard to the four curves is that the various clearance spaces are based on various initial pressures and temperatures.

In figures 155 to 158 diagrams are given showing for the four cases the ideal steam consumption per I. H. P. with compressions of nil, 20%, 40% and 90% (una-flow) with various cut-offs. These diagrams show the advantage which the una-flow engine with exhaust to the atmosphere possesses with earlier cut-offs. At later cut-offs the una-flow engine is not so favourable as the ordinary engine in which a low compression is obtained with late cut-offs. The effect of different compressions on the steam consumption may also be seen from figures 155 to 158.

In the diagram shown in figure 159, the ideal steam consumption values for different mean pressures are drawn for different values of the clearance with initial

Superheated steam 13 at. abs. 300° C Condensing (0.1 at. abs.) Clearance space 2%

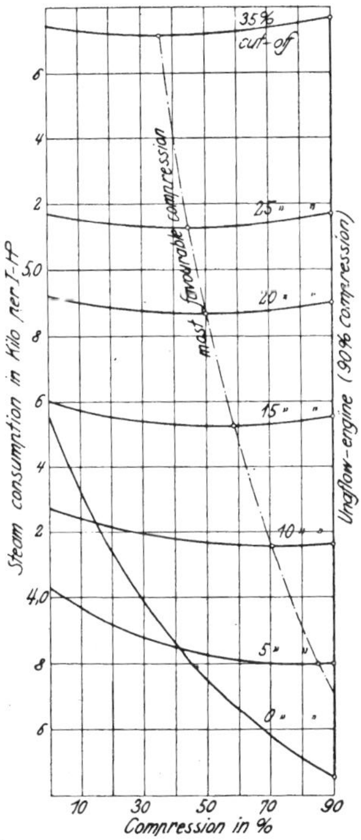

Fig. 160.

Superheated steam 13 at. abs. 300° C Condensing (0.1 at. abs.) Clearance space 5%

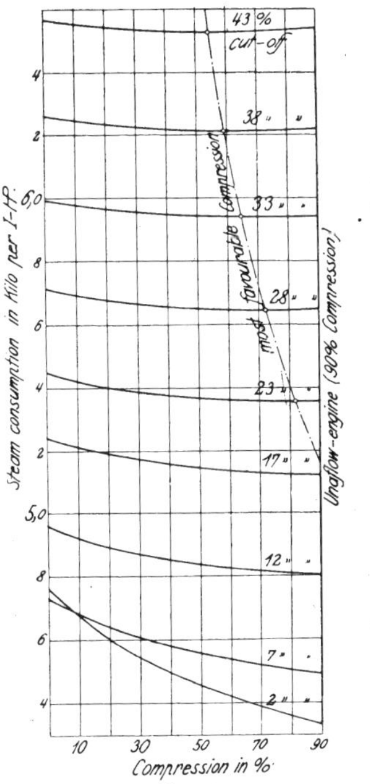

Fig. 161.

pressure of 13 atmos. abs. and initial temperature of 300° C, and exhaust to the atmosphere. The ideal steam consumption values are derived in the manner mentioned above. From this diagram it is interesting to note that the steam consumption for a mean pressure of 4 kgs per sq. cm, increases by approximately half a kilogram by increasing the clearance from 0% to 10%. The increase in the steam consumption at higher mean pressures is still greater as will be seen from the steeper inclination of the curves. These diagrams all clearly point to the fact *that the designer has every good reason for retaining the clearance space as small as possible.*

The conclusion arrived at from the above investigation is that the una-flow locomotive or other non-condensing una-flow engines, with their constant compression for 90% of the stroke, are specially advisable when using saturated steam and an early cut-off, the saturated steam allowing smaller clearance space than superheated steam.

When a supplementary compression space is used, as described in a previous chapter the superiority of the una-flow engine is unquestionable even at low initial pressures and high initial temperatures. At high initial pressures and low initial temperatures which render small clearance volumes possible the advantage is still more marked.

The diagrams for the steam consumptions per I. H. P. in an ideal condensing engine are shown in figures 160—161 for various cut-offs and compressions. In figure 160 the steam pressure is 13 atmos. absolute; temperature 300° C; condenser pressure 0·1 atmos. and clearance volume 2%. In figure 161, the steam pressure is 13 atmos. absolute; temperature 300° C, condenser pressure 0·1 atmos. and the clearance is 5%. The right hand ordinate represents 90% compression and the points where the curves intercept give the ideal steam consumption for this compression at the various cut-offs marked on the curves. As in the previous group of diagrams, the line of most favourable compressions is shown in dotted lines and is marked "most favourable compression". Referring to figure 160, it will be seen that the una-flow (90%) compression coincides with the most favourable compression possible up to cut-off at 3% of the stroke. With cut-off at 10% the best compression would be about 70%. At the same time it will be noticed that even with cut-off at 10% the increase in steam consumption by working with a 90% compression is negligible.

As the cut-off in a una-flow condensing engine is normally at about 8% of the stroke, it is evident that with condenser pressure at 0·1 atmos,, and a clearance of 2% the una-flow compression of 90% is approximately accurate. With a better condenser vacuum the approximation would be even more close. If all the features mentioned in the first chapter of this book are properly attended to, the most favourable compression will be obtained. These features consist in the use of a first class condensing apparatus and a close connection of the exhaust belt to the condenser, so that the exhaust belt opens directly through a wide and short connection into the condensing chamber.

All things considered it would appear from the curves discussed that a 90% compression is the closest approximation to the ideal. Thus even with power fluctuations of 50% above and below the normal, the departure from the most favourable steam consumption by working with a 90% compression is infinitesimally small. In the una-flow engine the constant compression is therefore quite admissible as for 50% overload the cut-off of 8% is only increased to about 12%. Thus on referring to figure 161 it will be seen that a 50% overload only involves an increase of 0.46% on the steam consumption per I. H. P. as compared with the ideal consumption for the best compression corresponding to this overload. This investigation proves conclusively that the objections so frequently raised against the high compression used in the una-flow engine are entirely unfounded. *With a 2% clearance, cut-off at 8%, condenser pressure 0.07 atmos absolute, the most favourable compression is 90% and these conditions may be taken as normal for the una-flow engine.* This statement assumes adiabatic compression and expansion, as has been found to exist in the una-flow engine. The same assumption is made in arriving at the diagrams shown in figures 149—164.

Superheated steam 13 at. abs. 300° C
Condensing (0.1 at. abs.)

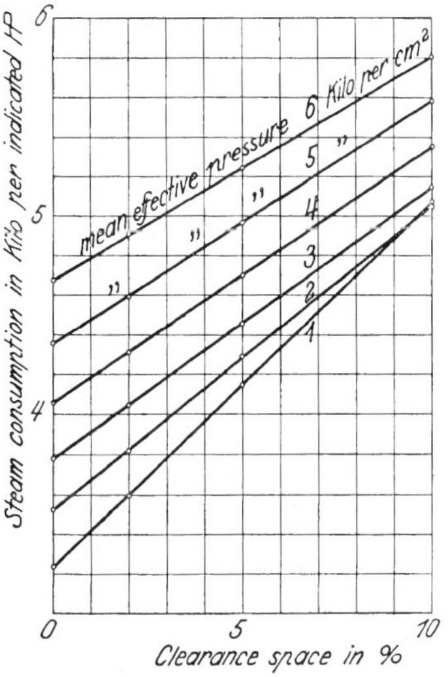

Fig. 162.

Superheated steam 13 at. abs. 300° C
Non condensing

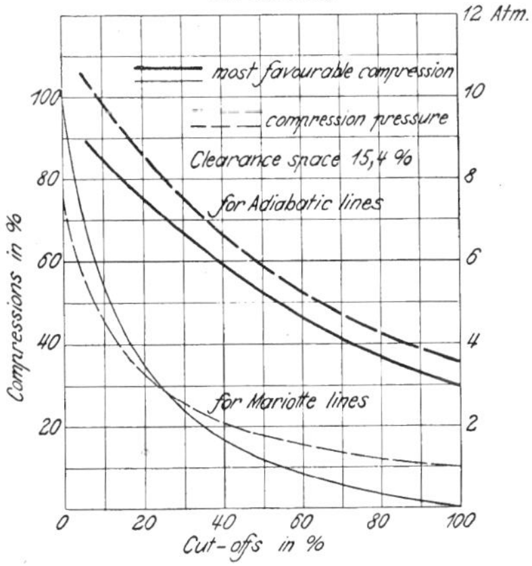

Fig. 163.

With a 5% clearance, steam pressure 13 atmos. absolute steam temperature 300° C and a condenser pressure of 0.1 atmos. absolute, the best compression exceeds 90% with all cut-offs up to about 18% of the stroke. With cut-off later than 18% of the stroke, the best compression is less than 90%. With cut-off at 43% the best compression is 54%.

Figures 150—161 show also that in ordinary single stage counterflow condensing engines where the exhaust valve is large and the clearance consequently increased, the compression is always too small. This also applies, though not in the same degree, to multiple expansion engines.

In figure 162, the ideal steam consumption in kilos per I. H. P. are given for various mean pressures and clearances. The initial steam pressure is 13atmos.

absolute, the temperature 300°C and the condenser pressure is taken at 0·1 atmos. absolute. It is interesting to note from this diagram that with a mean pressure of 1 kilo per sq.cm. the ideal steam consumption increases by nearly 1 kilo with a 5% increase in the clearance.

The clearance losses are relatively as well as absolutely reduced at higher mean pressures.

In taxing the values of the various compressions, consideration ought to be given to the reduced cylinder dimensions and the increased mean pressures, obtained by employing compressions below the ideals.

Superheated steam 13 at. abs. 300° C
Condensing (0.07 at. abs.)

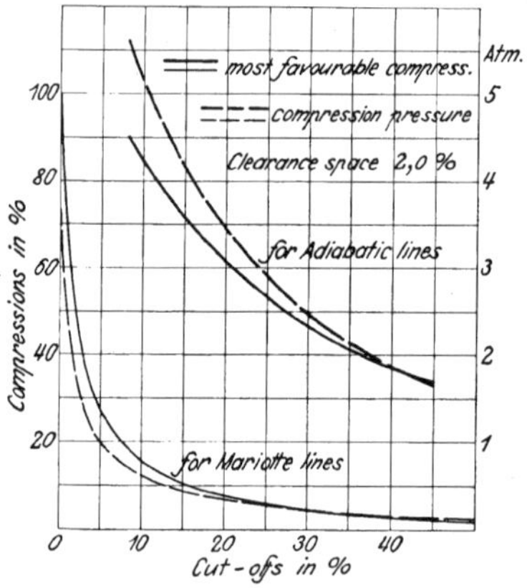

Fig. 164.

Closer investigation points to the conclusion that it is preferable to overcome the losses due to the clearances by reducing the clearance rather than tampering with the compression.

In the case of the una-flow engine it is much better to overcome the clearance losses by reduction of the clearance volume up to the limit when the end pressure of compression is equal to the initial pressure.

The economy effected in this way is much greater than the economy effected by using a larger clearance and obtaining a closer approximation to the most favourable compression.

Another point to bear in mind in considering these diagrams is that the adiabatic compression is greater than that given by the Mariotte line. In this connection attention might be drawn to figures 163—164. In figure 163 the steam is taken at 13 atmos. absolute and 300°C temperature, exhaust is to the atmosphere and a clearance of 15·4% is assumed. In figure 164 the steam is also at 13 atmos. absolute and 300° C temperature whilst the condenser pressure is taken at 0·07 atmos. absolute and the clearance is taken at 2%. In both diagrams the amount of the most favourable compression is shown in relation to various cut-offs in one case for the Mariotte line and in the other for the pure adiabatic. The corresponding end compressions are also inserted.

From figure 164 it will be clearly seen that 90% compression is correct with cut-off at 8%.

In addition to the considerations above regarding the steam losses entailed by the clearances it must be recollected that the clearance and compression must be proportioned to give efficient cushioning.

Chapter X.

The una-flow portable engine.

The una-flow engine is very well adapted for use as a portable engine. In the case of a portable engine, it is necessary that the engine be simple, light, cheap and economical as regards the use of steam. A small weight is specially necessary in the case of travelling portable engines. All these requirements are fully met by the una-flow steam engine.

Figures 165 to 167 show a portable engine built by the Maschinenfabrik Badenia vorm. Wm. Platz Söhne A.-G., Weinheim, Baden.

As compared with the ordinary portable engine employing tandem cylinders, one cylinder with all its accessories is completely dispensed with. Compared with the ordinary compound portable engine using lift valves, an entire single cylinder engine is saved, and on the remaining engine cylinder two valves are dispensed with. The una-flow steam engine only requires two valves, whilst the ordinary compound engine using lift valves requires eight. It is a point of considerable importance that the exhaust valves are omitted. The cylinder may be supported directly on the boiler and the inlet valve arranged vertically in the covers (see figure 168). The valves are operated from an eccentric arranged alongside the fly wheel and controlled by a fly wheel governor. The motion is transmitted to the valves in the same way as described above as applied to the ordinary form of una-flow engine. The engine frame is forked and the shaft is cranked between the two bearings. Two overhung fly wheels are arranged on the ends of these shafts. One of the fly wheels carries the governor and the other the driving gear (figure 169). The design is so arranged however, that the governor wheel may change places with the driving wheel and the condenser may be arranged on the right or the left hand side. All the parts are so arranged that they may be set up in either way. This is a very important feature where the design is

Fig. 165.

Fig. 166.

Fig. 167.

Fig. 168.

Fig. 169.

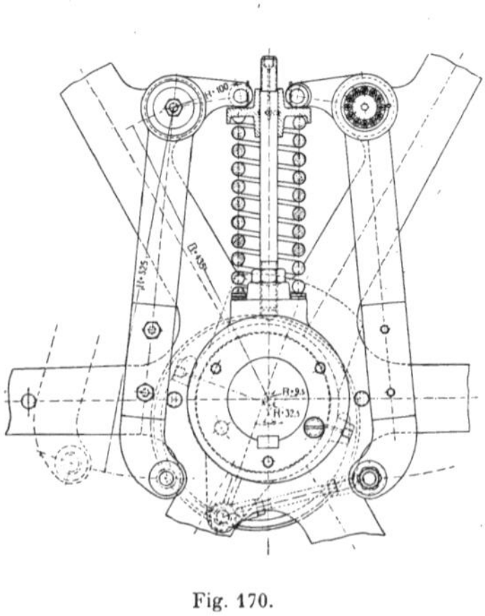

Fig. 170.

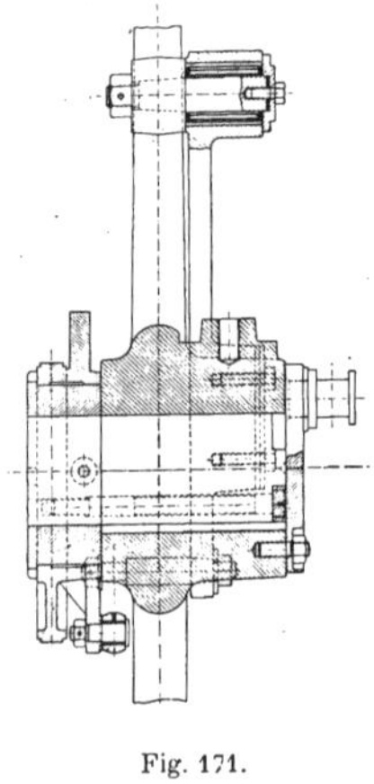

Fig. 171.

to be applied to a stock article. The vertical condenser pump is driven from a pin fixed on the fly wheel. These portable engines may be arranged for working as condensing engines or with exhaust to the atmosphere. By using additional clearance spaces, which are arranged in the cover and may be cut out when not required, the change over may be made at any time from condensing to atmospheric exhaust. The balancing of the reciprocating masses is effected by means of weights on the crank and fly wheels. The vertical centrifugal forces may be balanced by the moving parts of the condenser and feed pump. To give the best balancing, the feed pump should be arranged on the opposite side of the boiler to the condenser and adequate compensating weights provided.

The governor (figures 170 and 171), employed by the Badenia Company in Weinheim for their portable engine, is provided with two pendulum levers pivoted on two pins fixed on the fly wheel. These pendulum levers are of bell-crank form and have rollers mounted on their short ends. The rollers both act on a main pressure plate which is held up by a common spring. A pin is provided on the long end of each of the pendulum levers, but only one of these pins is connected to the shifting eccentric. Both levers, however co-operate to shift the eccentric. The free lever, not being handicapped by the eccentric, exercises more force on the pressure plate than the coupled lever and thereby relieves the other lever so as to enable it to exercise a greater force for the purpose of shifting the eccentric.

The shifting eccentric moves on a base eccentric, which is fixed to the shaft. This shaft eccentric is provided with holes and may be shifted through $180^0-2\delta$ and fixed in the shifted position (δ being the angle of advance). The shifting eccentric may then be swung round and the coupling rod joined up between the shifting eccentric and the other pendulum lever. The governor and valve gear are then set for the opposite direction of rotation. This, it will be seen, is done without altering the construction of any of the parts. This governor is only of use for high speeds, as the weight of the pendulum levers would be detrimental at low speeds on account of their natural oscillations.

The motion is transmitted from the adjustable eccentric to the valves in the manner described above, that is to say, a rocking lever is introduced and the valve rod is provided with a roller located in a groove. It is preferable to cast the rock shaft and rocking lever in one piece of steel.

Figure 172 shows the arrangement of a 100 HP. portable engine built by the Badenia Company. The cylinder is bolted down to the top of the boiler and is rigid with the forked frame. The bearings sit on flattened bars, (an arrangement peculiar to the Badenia Company) which permit free expansion of the boiler and ensure correct alignment of the shaft under all circumstances. In consequence of this excellent construction, the bearing difficulties frequently found in portable engines are absolutely removed. Each bearing really comprises two bearings. The inner of the two bearings comes close up to the crank shaft and takes the steam pressure forces whilst the outer of the two bearings comes close up to the fly wheel. In the centre of this double bearing, there is arranged a chain lubricating device. One of the fly wheels is provided with teeth for starting, the other

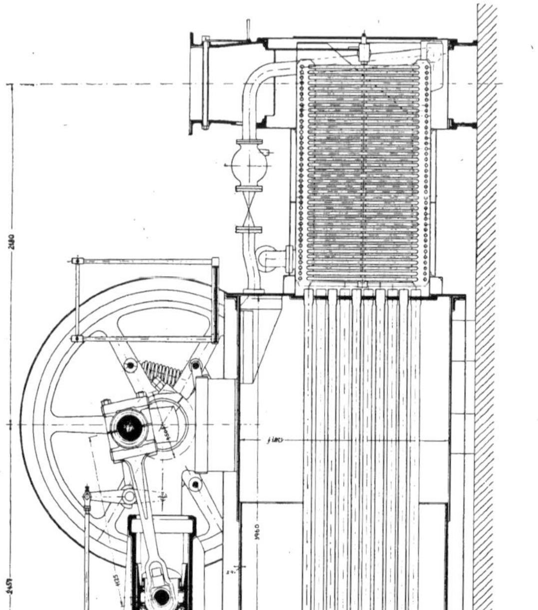

Fig. 172.

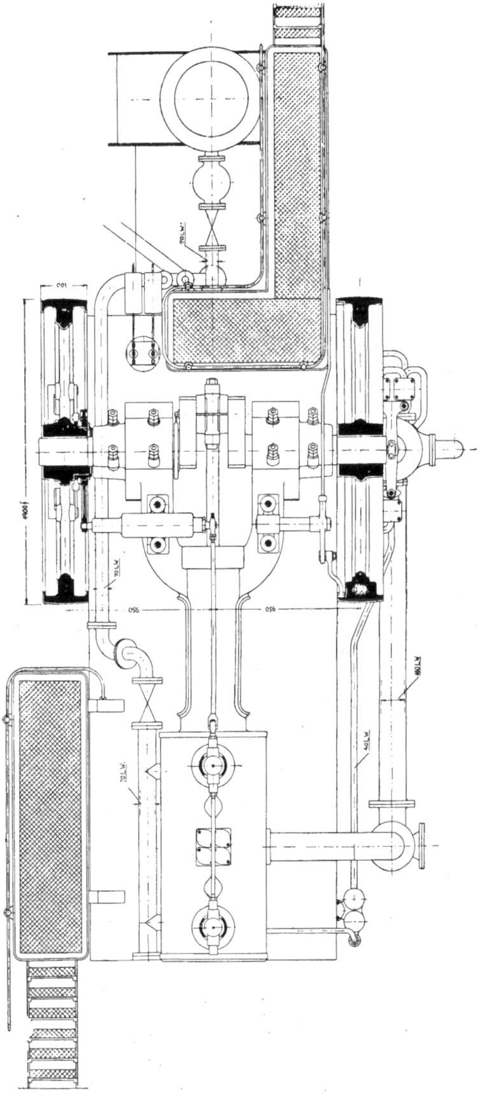

Fig. 172 a.

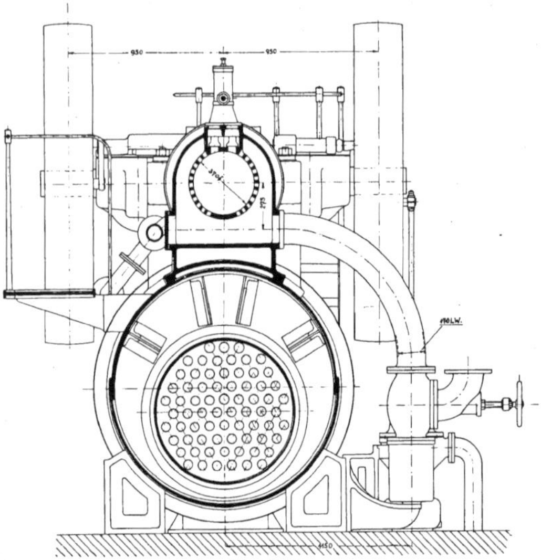

Fig. 172 b.

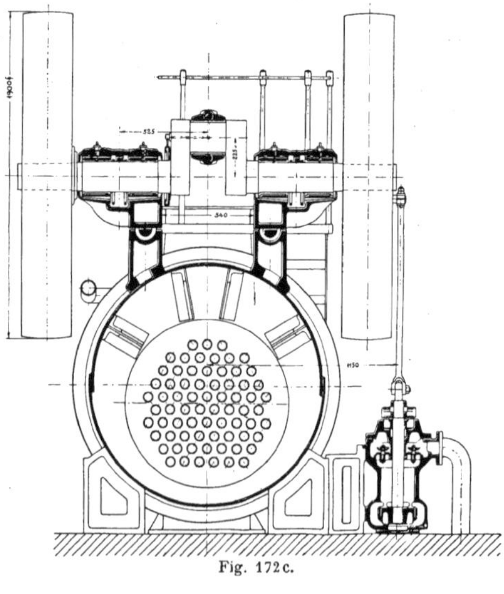

Fig. 172 c.

carries the shaft governor described above. The cylinder is bolted down to the boiler about its middle where the exhaust belt is arranged. Two openings are provided, leading from the exhaust belt on each side, so as to enable the condenser to be arranged at one side or the other. All the parts of the engine, including the fly wheels, frame, cylinder, condenser, valve gear etc., are so arranged that the condenser may be located on one or other side, or the engine may work with exhaust to the atmosphere or with a condenser and that the direction of rotation may be changed to suit the requirements of the purchaser or user. The principal features in favour of the una-flow portable engine reside in the central support for the cylinder on the boiler, the arrangement of the valve gear on the cylinder, and the well balanced connection between the shaft bearings and the cylinders consequent upon the use of the forked frame, and also the freedom of expansion between the frame and the boiler. To all this has to be added the simplicity of the entire construction. Attention should also be drawn to the superheater employed by the Maschinenfabrik Badenia (see figures 173 and 174).

The principal features of this construction are the small throttling loss in the superheater, the efficiency of its heating surface and the accessibility of the superheater and boiler tubes. The superheater is arranged at the smoke box end of the boiler in such a manner that the part of the boiler end plate pierced by the smoke tubes remains free (figure 173).

In figure 174, the same result is obtained by employing a zig-zag superheater tube running between the smoke tubes. This construction, as will be readily understood, enables the smoke tubes to be easily cleaned.

Figure 175, shows a portable una-flow engine built by Messrs. Robey & Co. Ltd. of Lincoln, England.

Figures 176 and 177, show a una-flow portable engine built by the Erste Brünner Maschinenfabrik Gesellschaft.

Figures 178 to 180, show a general arrangement as well as the details of the cylinder and governor of a portable una-flow engine for agricultural purposes built by the Kolomnaer Maschinenbau-A.-G. of Kolomna, near Moscow.

This portable type of engine must necessarily be simple in construction, cheap and still efficient in performing the duty required of it. To this end the unbalanced inlet valves are arranged with their spindles horizontal and between the ends of the spindle an arm of a three horned lever is located. Two of the horns in this lever co-operate with an oscillating arm on a counter shaft rocked from the eccentric. The arm on the counter shaft is provided with a roller at its end, which moves idly over the intermediate lever for the greater part of its stroke, but at the end of its oscillatory movement, the roller knocks the intermediate lever first one way, and then the other. The eccentric is controlled by a shaft governor. The entire valve mechanism is located in a chest cast on the exhaust belt, which chest may be filled with oil. The cover of this chest is so formed that the funnel of the boiler may be tilted back to rest thereon.

The valves in this small type of engine are spring loaded and unbalanced and are formed in one piece with the spindle. In larger engines the valves would be partially balanced and resilient. The horizontal arrangement of the valves

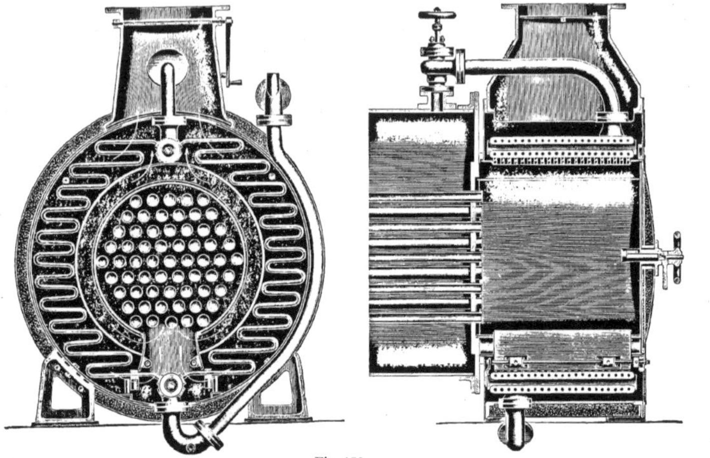

Fig. 173.

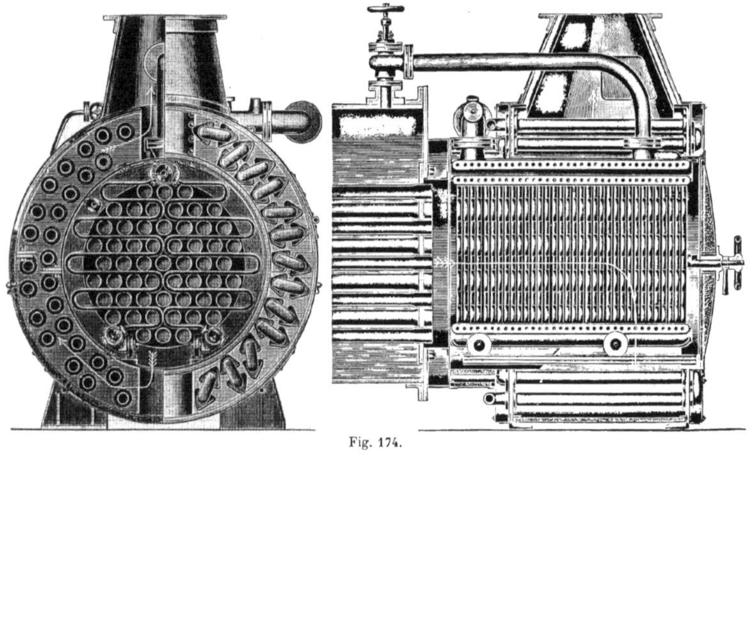

Fig. 174.

Fig. 175.

Fig. 176.

is adopted in order to simplify the construction. For the same end the front cover was cast in one piece with the cylinder.

This portable engine is designed to work with saturated steam at a pressure of 10 atmos.

In accordance with the information obtained from prior tests, the cover and the cylinder walls near the ends were heated. A neutral unheated zone is left

Fig. 177.

between the central exhaust belt, which acts as a cooling jacket, and the hot end jackets on the cylindrical walls.

The cylinder rests on a casting in which the stop valve is arranged (Fig. 181). This stop valve is adapted to admit steam simultaneously to both ends of the cylinder as shown. The cross section of this casting is such that any water of condensation deposited in the same may flow back to the boiler. The piston is built with two end pieces and is recessed at its end, so as to accomodate the necessary clearance space.

Figures 182 to 187 show the guides, stuffing boxes, crosshead, connecting rod, shaft bearing and crank shaft of this portable engine.

The extremely simple construction of the shaft governor is especially interesting. The controlling spring is anchored at one end to the flywheel and its other

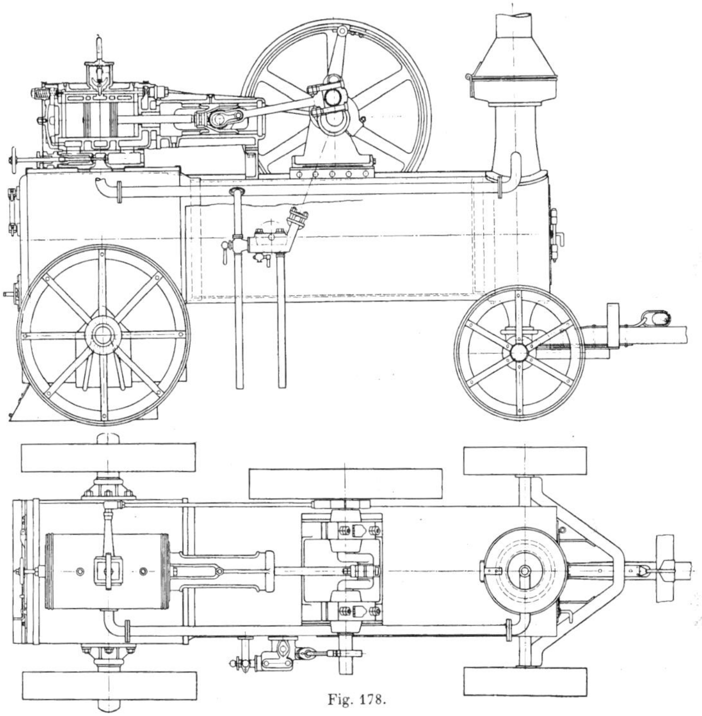

Fig. 178.

end is fixed to a centrally pivoted lever which is adapted to rotate on ball bearings about a pin fixed on one of the arms of the flywheel.

The opposite end of the centrally pivoted lever engages the driving eccentric. The spring performs the double function of supplying the centrifugal weight and the spring force. No spring forces or centrifugal forces are conducted through pins or bearings. Only the small shifting force of the eccentric is carried by the pivoted lever. This construction provides a very simple and cheap shaft governor in which, apart from the eccentric itself, no lubricating difficulties are introduced.

By placing the spring on the opposite side of the lever, the governor is arranged to act for the opposite direction of rotation. A free eccentric, such as is used in this construction, is only permissible for small powers when working at high speeds and with a strong construction of governor. In other cases, it is advisable to employ a self-locking eccentric which moves on a basic fixed eccentric, or to use a form of eccentric in which the reactionary forces act at small leverage.

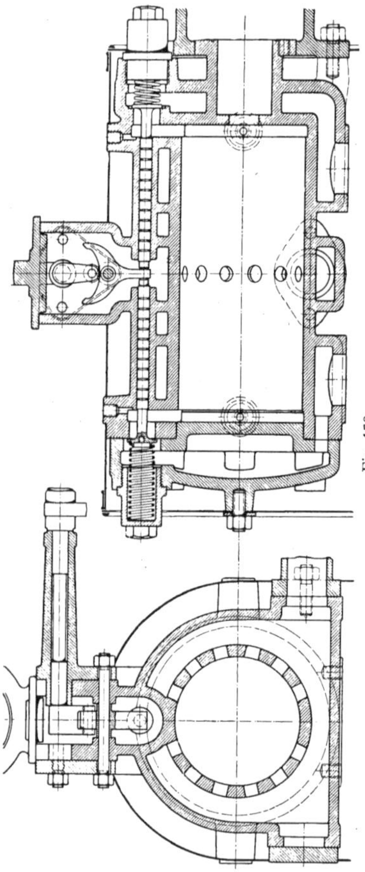

Fig. 179.

The boiler is provided with a large furnace chamber so that straw and wood may be used for firing.

The general arrangement drawing is shown in figure 178, and the picture shown in figure 188. An examination of this drawing will show that the construction is simple and well suited to the nature of the work. At the same time experience and stringent tests have shown that this construction yields excellently economical results as regards steam consumption.

The construction shown in Figure 178 has been improved in some details and these improvements are shown in figures 189 to 192. The ends of the cylinder and the cover are jacketted in the manner described above as suitable for saturated steam. These jackets form a kind of steam receiver in which a considerable volume of steam may be accommodated. The end jackets are connected directly to the boiler by flanged connections and the steam flows directly through two openings in the boiler shell into the steam receivers at the end of the cylinder. Owing to the considerable volume of the steam receivers the velocity of flow is small in the direction of the inlet valves at the upper

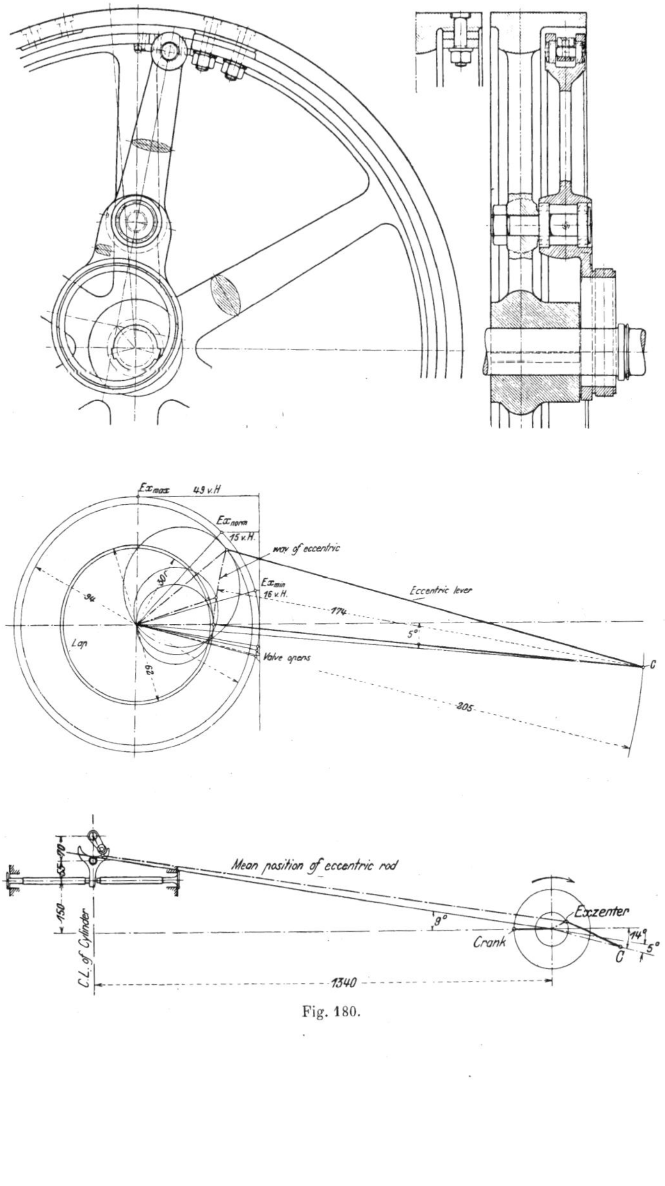
Ex max
49 v.H
Ex norm
15 v.H.
way of eccentric
Ex min
16 v.H.
Eccentric lever
30°
94
174
5°
Lap
29
Valve opens
C
205
70
55
150
Mean position of eccentric rod
Exzenter
9°
Crank
14°
5°
C
C.L. of Cylinder
1340

Fig. 180.

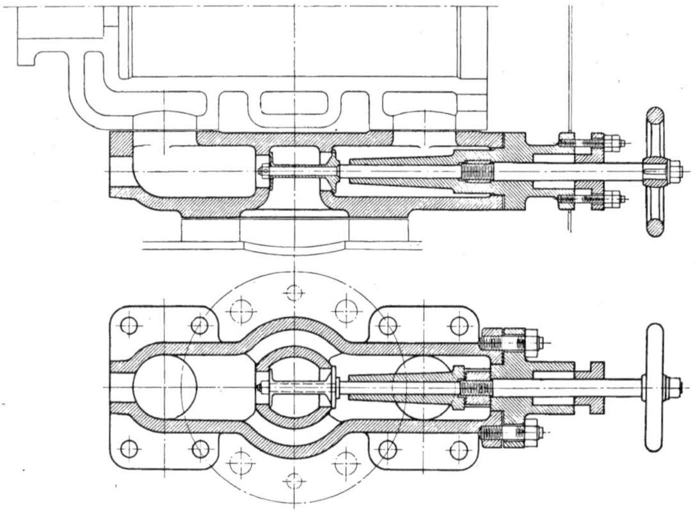

Fig. 181.

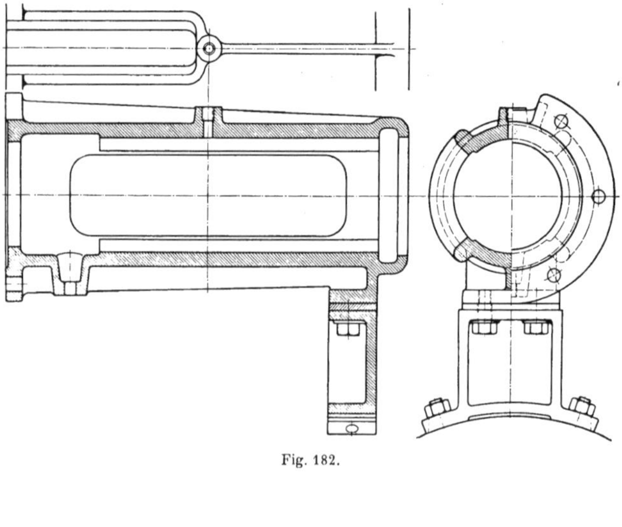

Fig. 182.

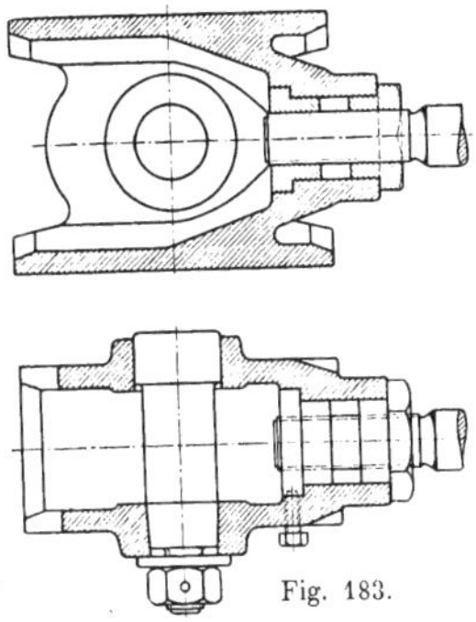

Fig. 183.

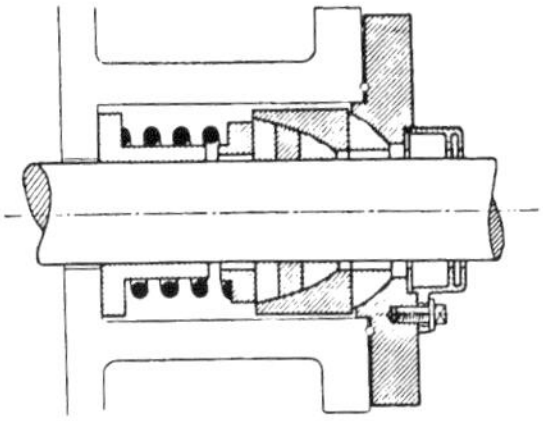

Fig. 184.

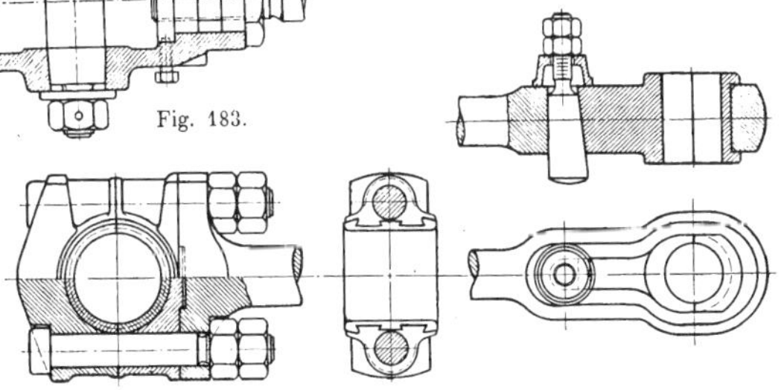

Fig. 185.

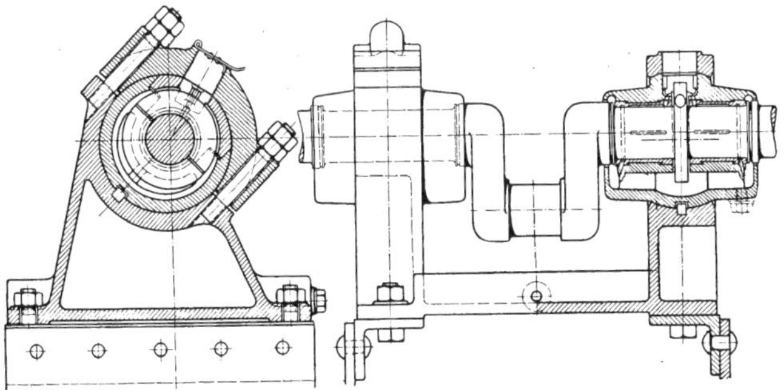

Fig. 186.

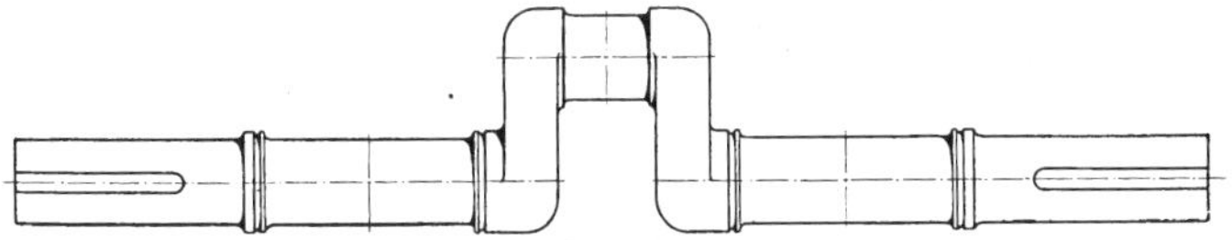

Fig. 187.

part of the cylinder. With this small steam velocity the water carried in suspension has ample time to be separated out and passed back to the boiler. This also applies to any water of condensation formed in the jacket itself, This design, therefore, provides, in addition to efficient end jacketting or heating of the cylinder, considerable advantage in its action as a steam drier and the engine works more economically owing to the dry state of the working

Fig. 188.

steam on its entrance to the cylinder. The inlet valves are enclosed in a separate compartment which may be closed off so as to prevent the passage of steam from the combined steam receiver and end jacket into the valve compartment. The diagram taken from this engine shows that there is a considerable improvement in the admission line due to the action of the receiver which is similar to that of an air vessel in a pump.

Another advantage which this construction has over that shown in Figure 178 is that the centre of the cylinder is brought nearer to the boiler

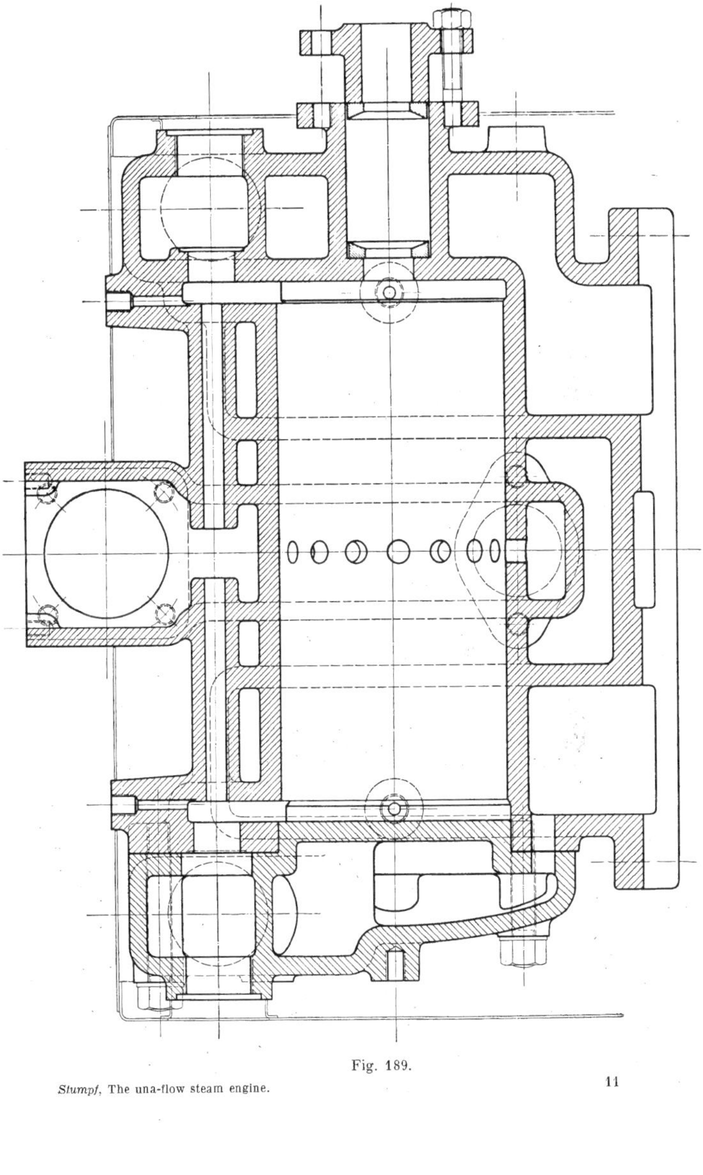

Fig. 189.

and the casting interposed between the boiler and the cylinder, and which contained the stop valve, is omitted. The cylinder is automatically heated up before starting so that no delay or difficulties occur at such times. It has proved advantageous in practice to provide the exhaust ports all round the cylinder and especially at the lowest part in order to allow any water to pass off freely.

The following is a translation of the reports on tests made by independent investigators of a semi-portable engine built by the Maschinenfabrik Badenia.

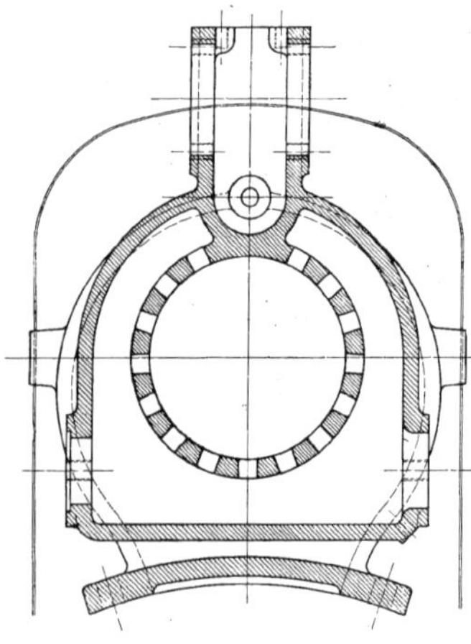

Fig. 190.

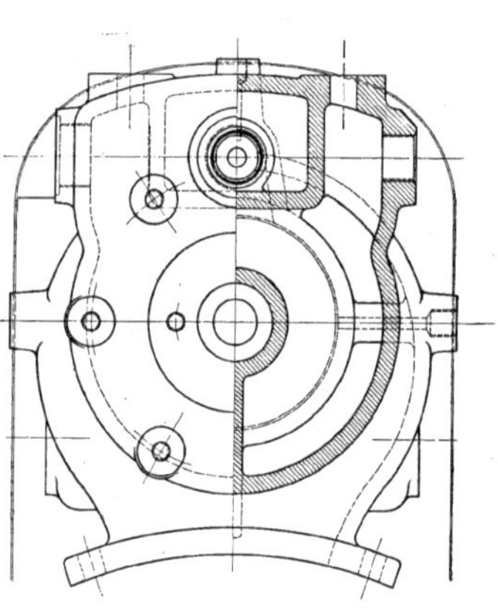

Fig. 191.

Translation.

Württembergischer Dampfkesselrevisions-Verein, Stuttgart (Boiler Inspectors Society of Wurtenburg) Department II.

Stuttgart, 14th October 1910.

Evaporation and brake tests made on a una-flow portable engine working with superheated steam [shop No. 3720], on the 27th, 28th and 29th September 1910 in the Maschinenfabrik Badenia, vorm. Wm. Platz Söhne, A.-G., Weinheim, Badenia.

Purpose of the Test.

The test was for the purpose of determining whether the contracted guarantees were fulfilled.

The following items were to be measured

1. The coal consumption per hour per Brake horse power when working normally with an output of 220 to 240 B.H.P.
2. The steam consumption per B.H.P. at normal load.
3. The maximum continuous output of the portable engine.
4. The temporary overload.

Guarantees.

1. A coal consumption of 0·59 to 0·61 kg for each B.H.P. at normal load of 220 to 240 B.H.P., the fuel used being coal with a calorific value of at least 7500 heat units (metrical).

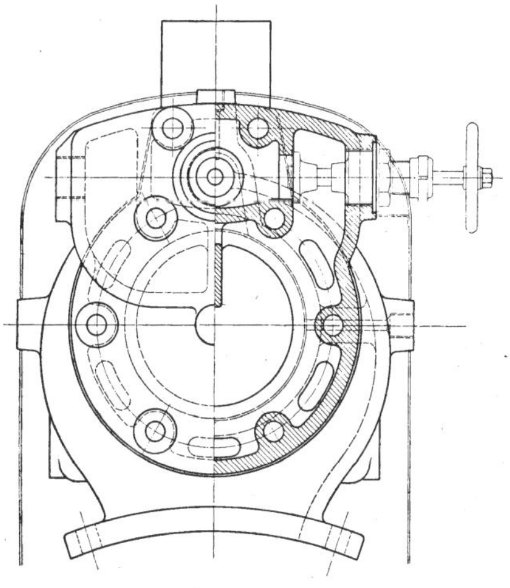

Fig. 192.

2. A steam consumption of 4·9 to 5·0 kg per hour per B.H.P. at normal load.
3. A maximum continuous load of 310 B.H.P.
4. A temporary overload of 350 B.H.P.
5. Smooth silent running of the engine without substantial longitudinal or transverse vibrations.

Construction and Design.

a) The Boiler.

This is a horizontal firebox boiler with horizontal fire tubes opening into the smoke box. The corrugated firebox together with the fire tubes are removable. The superheater is arranged in the smoke box and consists of flat wrought iron tubular coils.

The furnace gas feed water heater is made of straight wrought iron tubes connected in series by cast iron bends at their ends. This heater is arranged in a flue surrounding the smoke box.

The hot gases pass from the fire box, through the boiler tubes, sweep over the superheater arranged in the smoke box, and are then led through the flue surrounding the superheater and containing the feed water heater. From this preheater flue, the furnace gases escape to the chimney.

Particulars.

a) The heating surface of the boiler . . . . 70·6 sq.m.
Normal Grate Area . . . . 1·2 »
Grate area during the duration tests . . . . 0·95 »
The ratio of Grate area to Heating surface during the duration test . . . . 1 : 74·3.
Heating surface of the superheater . . . . 38·5 sq.m.
Heating surface of the feed water heater . . . . 13·5 »
The maximum Steam pressure . . . . 12 atmos.

b) *Engine*

The engine is a horizontal una-flow engine arranged on the boiler and provided with lift valve gear. Between the cylinder and the condenser, an exhaust steam preheater is arranged. The feed water to the boiler is first pumped through the exhaust steam heater and then through the furnace gas preheater.

Dimensions.

The cylinder diameter . . . . 525 mm.
Piston Stroke . . . . 600 »
Diameter of the piston rod at front . . . . . . 94·8 »

Conditions of the Test.

The tests were carried out in accordance with the normal conditions laid down for energy tests.

The tests were not made on the future working position of the engine but in the builders' workshop.

The necessary draught was obtained from a suction ventilator, which drew the gases into an iron chimney.

In accordance with the requirements set down for energy tests, the hourly coal and steam consumption at normal load of the engine was taken on two occasions on successive days.

Test I, was made on the 27th September and lasted from 8.25 a. m. to 4.29 p. m., that is, eight hours and four minutes.

Test II, was made on 28th September, began at 8.24 a. m. and ended at 4.28 p. m. and lasted without intermission for eight hours, four minutes. The coal and feed water were weighed. The steam pressure, temperature and draught readings were taken every quarter of an hour, as also the relative amounts of carbonic acid and oxygen in the furnace gases leaving the preheater.

The brakes used had two wrought iron brake bands provided with wooden pads and these brakes were laid around both brake discs. The weights for the brake bands were slung on rods which engaged in eyes on the brake bands.

The revolutions of the engine were taken by a tachometer operated from the engine.

Indicator diagrams were taken at convenient intervals to determine the cut-off.

During both tests, the fuel used was coal briquettes, presumably from the Oranje-Nassau pit.

A sample taken during the test was examined in the Engineers' laboratory of the Technical High School in Stuttgart to determine its calorific value. The calorific value found was 7592 heat units, whilst the moisture contained was roughly 1.95% and the ash 7.24%.

On the day previous to the test, that is on the 26th September, a preliminary test was made in order to find out whether or not the guaranteed maximum continuous load and the maximum temporary load were reached.

The brake test for determining the maximum continuous load lasted uninterruptedly for one hour.

The test for the temporary maximum overload was made immediately thereafter and lasted without interruption for thirty minutes.

Results of the Tests.

The results of the evaporation tests I and II, are given in the accompanying table A. The results for the brake tests I and II are given in the accompanying table B. The results of the brake tests for the maximum duration load and the temporary maximum overload are given in the accompanying table C.

As seen from the results of the brake test in Table B, the output during Test I, was 234.85 B.H.P. per hour, with a coal consumption of 0.587 kg. (coal briquettes) and 4.85 kg. of steam for each B.H.P.

In the case of Test II, the output was 235.53 B.H.P., with a coal consumption of 579 kg. per hour per B.H.P., and a steam consumption of 4.84 kg. per B.H.P. hour.

The average results of these two tests, which practically agreed, are as follows:

1. The hourly coal consumption with an output of 235.19 B.H.P. averaged 0·583 kg. per B.H.P. with fuel having a calorific value of 7592 heat units [Centigrade]. Calculating this coal consumption for a fuel with a calorific value of 7500 heat units which is the value referred to in the guarantees, the hourly coal consumption per B.H.P. amounts to 0·590 kg., whilst the guarantee coal consumption was between 0·59 to 0·61 kg.

2. The hourly steam consumption per B.H.P. was 4·85 kg. at a boiler pressure of 12 atmos. and a temperature at the cylinder of 263° C. whilst the guaranteed steam consumption was 4.9 to 5 kg.

From the table C, the following is shown.

3. The maximum continuous load for this portable engine lasting over one hour with an average steam pressure of 12 atmos., and an average speed of 151.9 revs. per minute, was 321.5 B.H.P., the guaranteed maximum duration load being 310 B.H.P.

4. The maximum temporary overload for the engine lasting over half an hour with an average boiler pressure of 12 atmos. and an average speed at 148.1 revs. per minute was 354.8 B.H.P., the guaranteed maximum overload being 350 B.H.P.

Conclusion.

The results of the foregoing tests lead to the conclusion that all the guarantees given as regards power, coal and steam consumption for this portable engine are entirely fulfilled.

With regard to the guarantees for silent and smooth running of the engine, it is to be remarked that the engine at all loads ran quietly and smoothly without material longitudinal or transverse vibrations of the portable engine.

In this connection it should be further mentioned that the portable engine during the tests was placed on a foundation which was only designed for temporary purposes.

Report on Diagrams.

The diagrams taken during the test proved that the steam was properly distributed.

Table A.

Results of evaporation tests I and II on the 28th and 29th September 1910.

		I	II	Average
Test No		I	II	Average
Test Days		28. 9. 10	29. 9. 10	
Test Duration	mins	484	484	484
Feed water: total evaporation	kg	9200	9202	9201
Evaporation 1 hour (average)	kg	1140·5	1140·7	1140·6
» 1 » per sq. m. heating surface (average)	kg	16·16	16·16	16·16
Water temperature in suction well (average)	°C	13·4	13·1	13·25
» » on exit from exhaust steam heater (average)	°C	53·5	52·0	52·75
» » on exit from the furnace gas preheater (average)	°C	70·0	70·4	70·2
» » rise in steam preheater (average)	°C	40·1	38·9	39·5
» » rise in furnace gas preheater (average)	°C	16·5	18·4	17·45
» » rise in steam and furnace gas preheater (average)	°C	56·6	57·3	56·95

Fuel: total amount stoked . . . . kg	1113	1100	1106·5
» in 1 hour (average) . . . . kg	137·98	136·36	137·17
» in 1 hour per sq. m. grate area (average) . kg	145·2	143·54	144·37
Calorific value in dry state . . . . heat units	7648	7648	7648
Approximate moisture in original state . . . .%	1·95	1·95	1·95
Content of ash according to chemical analysis . . . . . .%	7·24	7·24	7·24
Calorific value in original state . . . . heat units	7592	7592	7592
Total residue on the grate . . . . kg	80	82	81
» » in % of fuel used . . . .%	7·0	7·4	7·2
Steam: pressure in the boiler (average) . . . . . kg/sq. cm	11·95	12	12
Temperature of saturated steam (average) . . . .°C	191	191	191
Temperature of superheated steam (average) on entrance to the engine . . . .°C	252	274	263
Average superheat . . . .°C	61	83	72
Heat necessary to generate saturated steam from water at 70° C . . . . heat units	598	598	598
Heat necessary to generate superheated steam from water at 70° C . . . . heat units	628·5	639·5	634
Air: Mean temperature, 4 m in front of the grate . . . .°C	23	24	23·5
Flue gases: Mean temperature in front of the superheater °C	399	426	412
Mean temperature behind the furnace gas pre-heater . . . .°C	184	208	196
Mean fall in temperature . . . .°C	215	218	216
Carbonic acid present at the end of the boiler (average). . . . .%	12·1	11·7	11·9
Oxygen present at the end of the boiler (average) %	7·8	8·1	7·9
Draught: in fire box in mm of water column . . . . . mm	6·3	6·2	6·2
at the end of the boiler . . . . mm	13·4	14·4	13·9
Fall from the end of the boiler to the combustion chamber (average in mm of water column) . . .	7·1	8·2	7·7
Evaporation: 1 kg of fuel generated steam . . . . kg	8·26	8·36	8·31
Cost of 10 000 kg of fuel at the locality where the tests were made . . . . Marks	209	209	209
Cost per 1000 kg of working steam . . . . Marks	2·538	2·508	2·515

Heat Distribution.

Test No	I	II	Average
Heat of fuel utilised in the boiler . . . .%	65·2	66·0	65.6
Heat of fuel utilised in the superheater . . . .%	3·3	4·5	
Heat of fuel utilised in the boiler and superheater . . . .%	68·5	70·5	
Heat of fuel utilised in exhaust steam and furnace gas pre-heaters . . . .%	6·2	6·3	
Total heat of fuel utilised . . . .%	74·7	76·8	
Heat lost to the furnace gases leaving the furnace gas feed water heater (about) . . . .%	8·6	10·0	
The remaining loss due to radiation, conduction and un-burned fuel (about) . . . .%	16·7	13·2	
Total	100·0	100·0	

Table B.

Results of Brake tests I and II 28th and 29th September 1910.

	I	II	Average
Test No	I	II	Average
Day of test	28. 9. 10	29. 9. 10	
Duration of test . . . mins	484	484	484
Steam consumpt. in the steam engine during test . . . kg	9200	9202	9201
Steam used in steam engine per hour (average) . . . . kg	1140·5	1140·7	1140·6
Steam pressure in boiler (average) . . . kg/sq. cm	11·95	12·0	12·0
» » on entrance to the cylinder taken from the indicator cards (average) . . . kg/sq. cm	11·4	11·3	11·35
Steam temperature on entrance to the cylinder (average) . °C	252	274	263
Fall in pressure between the boiler and engine (average) kg/sq. cm	0.55	0.7	0.62
Superheat of steam in front of the engine stop valve (average) . . . °C	61	83	72
Vacuum in Condenser (average) . . .	0·93	0·93	0·93
Temperature of the spraying water in the condenser (average) . . . °C	12·6	12·4	12·5
Temperature of water leaving condenser . . . °C	33	31	31·5
Cut-off in cylinder in % of stroke (average) . . . %	10·2	10·6	10·4
Revolutions per minute . . .	151·02	151·46	151·24
Brake: Weight (left) . . . kg	277	277	277
Weight (right) . . . kg	480	480	480
Fulcrum for the weight, left . . . mm	1612	1612	1612
» » » » right . . . mm	1390	1390	1390
Brake Power, left . . . H.P.	94·16	94·43	94·29
» » right . . . H.P.	140·69	141·10	140·90
Total Brake Power . . . H.P.	234·85	235·53	235·19
Coal consumpt. per B.H.P. hour . . . kg	0·587	0·579	0·583
» » per B.H.P. hour calculated for coal with a calorific value of 7500 Heat units, in accordance with guarantees . . . kg	0·595	0·586	0·590
Cost of fuel per B.H.P. hour . . . Pfg.	1·23	1·21	1·22
Steam consumpt. per B.H.P. hour . . . kg	4·86	4·84	4·85
Heat used per B.H.P. hour calculated on feed water at 0° C . . . heat units	3377	3421	3399
Heat contained in fuel usefully employed in engine . . . %	14·37	14·57	14·47

Table C.

Results of Brake tests III and IV on 26th September 1910.

	III	IV
Test No	III	IV
Day	26. 9. 10	26. 9. 10
Duration . . . mins	60	30
Steam pressure in boiler (average) . . . kg/sq. cm	12	12
Brake: Weight (left) . . . kg	527	651
Weight (right) . . . kg	480	480
Fulcrum for weight (left) . . . mm	1612	1612
» » » (right) . . . mm	1388	1388
Revolutions per minute (average) . . .	151·9	148·1
Brake Power . . . H.P.	321·5	354·8
Cut-off in percentages of the stroke (average taken from diagrams) . %	17	23

The late cut-off should be noticed. If the cut-off were earlier and the cylinders jacketted (which in this case was omitted), a very material reduction in the steam consumption could have been obtained. In spite of this however, an average steam consumption of 4·365 kgs per indicated H. P. hour was obtained assuming the mechanical efficiency to be 90%. The above result corresponds very closely to the 300 H.P. engine mentioned earlier.

Second Report.

Report of a brake, indicator and evaporation test made on the una-flow portable engine No. 3788, Marke Laubeone, without condensation and with superheat, built in 1910, for a pressure of 12 atmos.; builders, The Maschinenfabrik Badenia, A.-G. in Weinheim.

The Maschinenfabrik Badenia, sold the above portable engine under the following guarantees:

Coal Consumpt. per B.H.P. Hour at normal loads and with proper handling 0·85 to 0·88 kgs, coal used, having a heating value of at least 7500 heat units metrical and not having more than 5% ashes and slag.

Under the same conditions the steam consumption per B.H.P. hour is to be 7·3 : 7·5 kgs.

The portable engine is to give

Normal	120 B.H.P.
Maximum duration load	160 »
Maximum overload	185 »

The purpose of the test was to determine whether or not the above guarantees had been fulfilled.

The test was made under the conditions set down by the Verein Deutscher Ingenieure, the International Union of Boiler Inspection Societies and the Verein Deutscher Maschinenbau-Anstalten in 1899.

The boiler is a horizontal fire-box boiler with removable tubes, behind which a superheater is arranged.

The engine is a una-flow engine working with superheated steam, and having a valve gear of the Stumpf type.

The engine is arranged horizontally on the boiler.

The crank shaft carries a flywheel on the right hand and a brake-disc on the left.

The fuel and feed water, the latter being heated by the exhaust steam from the engine, were weighed as required. The temperature readings, the boiler pressure readings, and the measurements of the draught were taken every quarter of an hour. The speed of revolution was determined by a tachometer. Indicator diagrams were taken every half hour by two indicators manufactured by Dreyer, Rosenkranz & Droop.

The boiler dimensions necessary for the calculations, as well as the particulars of the fuel, feed water, temperature readings, and the general results of the test, are given below:

A. The boiler.

I. General particulars.

Heating surface of the boiler	47·00 sq. m.
» » of the superheater	23·40 »
Grate Area, normal	0·81 »
» » during test	0·70 »
Ratio of Grate Area to the heating surface:	
a) normal	1 : 58
b) during test	1 : 67·2
Ratio of the superheating surface to the boiler heating surface	1 : 2·01

A brick chimney was used with artificial draught.

II. Observations and results of the evaporation test.

Commencement of the test	9·26 a. m.
End of test	3·30 p. m.
Duration of the test	364 mins
Fuel.	
Kind and origin: briquettes Calorific value	7680 heat units
Total fuel used	600 kg
» » » in 1 hour	98·9 »
» » » » 1 » per sq. m. Grate Area	141·3 »
» » » » 1 » per sq. m. Heating surface	2·105 »
Ash.	
Total	43 kg
Percentage of fuel	7·17%
Feed Water.	
Total evaporation	5400 kg
» » in 1 hour	890 »
» » » 1 » per sq. m. Heating surface	18·95 »
Average temperatures of feed water	11·95° C
» » after passing through preheater	91·00° C
» » rise in preheater	79·05° C
Heat utilised in preheater:	
Total	426 870 heat units
Per kg of fuel	711 » »
Steam.	
Average pressure (above atmosphere)	12·0 kg per sq. cm
Temperature of saturated steam	191·0° C
» » superheated steam	275·0° C
Rise in temperature through superheater	84·0° C
Total heat for 1 kg of saturated steam	669 heat units
» » » 1 » » superheated steam	717 » »
Heat necessary to generate 1 kg of saturated steam	578 » »
Heat necessary to generate 1 kg of superheated steam	626 » »
Heat used in superheater per 1 kg of steam	48 » »
Total	259 200 » »
Per 1 kg of fuel	432 » »
Hot Gases.	
Average carbonic acid (CO_2) behind the superheater	11·5%
Average temperature of the hot gases:	
a) in front of the superheater	401° C
b) behind the superheater	255·5° C

Average temperature of the air for combustion	24° C		
Average draught in mm of water column, behind the superheater	13·0 mm		

Evaporation Results.

1 kg of fuel generated:

a) Steam (heat necessary to generate 1 kg = 626 heat units)	9·0 kg
b) Steam normal (heat necessary to generate 1 kg = 637 heat units)	8·85 »

Heat Balance:

Of the Heat contained in each kg of fuel there were used in the boiler	5202 heat units	67·8%
Used in the superheater	432 » »	5·6%
Total	5634 heat units	73·4%

Lost:

Heat lost to the chimney gases	998 heat units	13 0%
» » by radiation, conduction, residue and unburnt gases etc.	1048 » »	13·6%
Total	7680 heat units	100·0%

The exhaust steam preheater gave up 711 centigrade heat units per kg of fuel and this heat was utilised for heating the feed water 9·3%

B. Steam engine.

I. General Particulars.

Cylinder diameter	415 mm
Piston rod diameters:	
a) crank end	75 »
b) cover end	0 »
Effective piston surface:	
a) crank side	1308·47 sq. cm
b) cover side	1352·65 » »
Stroke	500 mm
Revolutions normal	180
Scale of indicator spring:	
a) crank side 1 kg =	3 mm
b) cover side 1 kg =	3 »
Constant in the formula $N = k \cdot n \cdot p$.	
a) crank side	0.1454
b) cover side	0.15025

II. Energy trials.

1. Normal output.

Radius of the brake	1425 mm
Weight on the brake	331 kg
Speed of Revolution	183·7
Total brake power	121·0 H.P.
Indicated Power:	
a) crank side	67·0 H.P.
b) cover side	76·8 »
Total indicated H.P.	143·8 H.P.
Mechanical efficiency $\frac{\text{B.H.P.}}{\text{I.H.P.}}$	84·6%
Steam consumpt. per B.H.P. hour	7·36 kg
Steam consumpt. per I.H.P. hour	6·19 »
Coal consumpt. per B.H.P. hour	0·817 »
Coal consumpt. per I.H.P. hour	0·687 »

2. Maximum continous Load.

Radius of the brake	1425 mm
Weight on the brake	456 kg
Revolutions per minute	180·3
Total B.H.P.	164·0 B.H.P.
I.H.P.:	
a) on the crank side	88·1 H.P.
b) on the cover side	100·2 »
Total I.H.P.	188·3 H.P.
Mechanical efficiency $\frac{B.H.P.}{I.H.P.}$	87·1%

3. Maximum load.

Radius of the brake	1425 mm
Weight on the brake	531 kg
Revs. per minute	178
Total brake power	188 H.P.
Indicated Power:	
a) crank side	97·6 H.P.
b) cover side	114·4 »
Total indicated H.P. of engine	212·0 H.P.
Mechanical efficiency $\frac{B.H.P.}{I.H.P.}$	88·7%

Conclusions.

As shown by the results of the test the guarantees given by the builders are fulfilled.

The engine ran without stopping during the time of the test. Its running was silent and uniform. The indicator diagrams were well shaped and showed efficient operation of the valves.

The test took place on the 30th March 1911, in the erection shop of the Maschinenfabrik Badenia, A.-G. in Weinheim and was conducted by Messrs. Freitag & Metzner of the undersigned company.

Mannheim, 12th April 1911.

Badische Gesellschaft zur Überwachung von Dampfkesseln e. V.

Der Oberingenieur

(sgnd) Pietzsch.

It was subsequently found that the steam leaked behind the outermost piston ring and pressed the latter during expansion and exhaust against the cylinder walls. This caused additional friction which had its effect on the mechanical efficiency.

The Maschinenfabrik Badenia, vorm. Wm Platz Söhne, A.-G. in Weinheim, Baden, has done particularly good service in the development and introduction of the una-flow portable steam engine. This firm has taken over the manufacture of the una-flow portable engine for Germany, the Erste Brünner Maschinenfabriks-Gesellschaft in Brünn for Austria Hungary, Messrs. Robey & Co. Ltd., of Lincoln for England and the Kolomnaer Maschinenbau-A.-G. of Kolomna for Russia.

Chapter XI.

The una-flow rolling mill engine.

The valve gear of a reversing rolling mill engine should be so arranged, that during starting and stopping the steam is throttled, whilst during the actual rolling process, normal diagrams with full steam pressure and the most-economical cut-off and expansion should be obtained. This requirement is met in the case of the engine shown in figure 193, and described below.

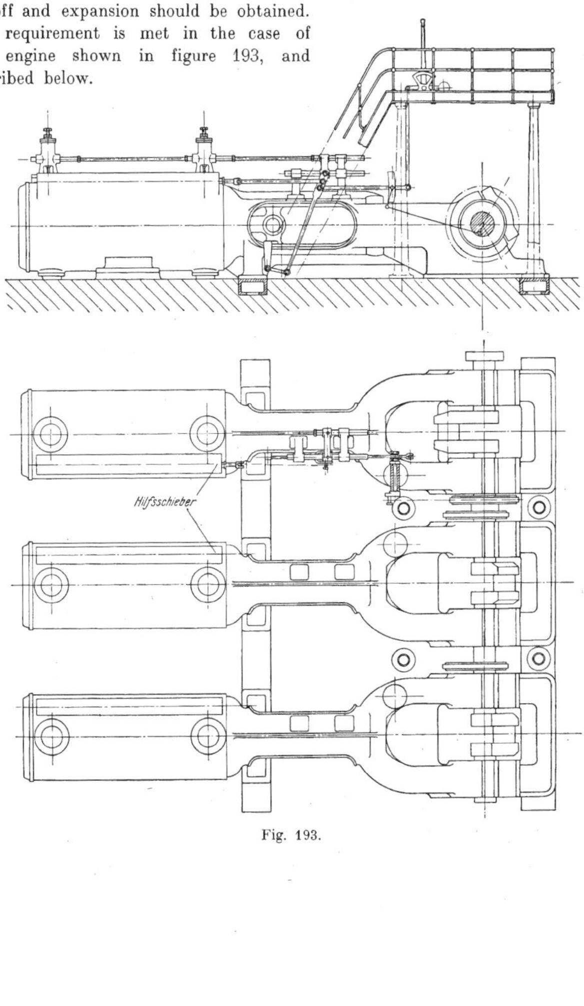

Fig. 193.

The valve gear is arranged to operate main inlet valves and also additional auxiliary inlet valves. The main inlet valves are designed to give full steam pressure, whilst the auxiliary valves are of small area and consequently throttle the incoming steam. The auxiliary inlet works with a late cut-off, whilst the

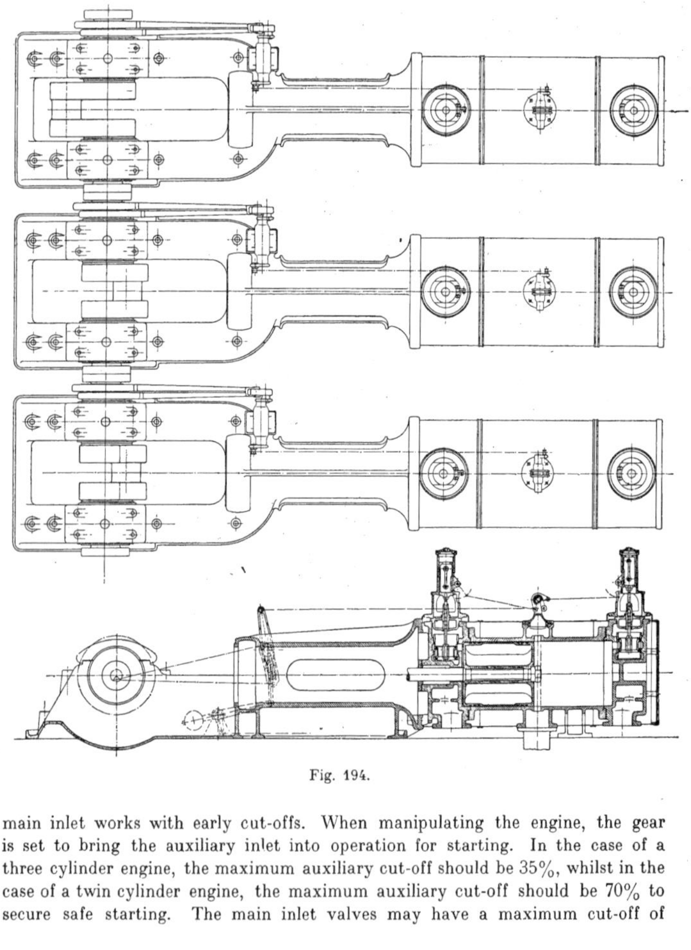

Fig. 194.

main inlet works with early cut-offs. When manipulating the engine, the gear is set to bring the auxiliary inlet into operation for starting. In the case of a three cylinder engine, the maximum auxiliary cut-off should be 35%, whilst in the case of a twin cylinder engine, the maximum auxiliary cut-off should be 70% to secure safe starting. The main inlet valves may have a maximum cut-off of half the above amounts. So long as the blank is not in the grip of the rolls, the engine cannot race on account of the considerable resistance to be overcome, even when running empty, and also on account of the small diagrams which always occur at high speeds. Such high speeds naturally increase the throttling in the auxiliary inlet valve, which is in itself designed to give a certain amount of throttling. As soon as the blank is gripped and the load on the engine is increased, the operator shifts the gear to bring the main inlet valves into operation, to meet the increased resistance. The speed is thus further increased and the rolling operation proceeds at the necessary rate. The action of the auxiliary inlet decreases and disappears gradually, until the diagram is ultimately determined only by the main inlet. During actual rolling there is therefore almost no throttling. Towards the end of the bite, the operator brings back the valve gear gradually to give a slow run out. The valve gear thus passes again through the positions which give throttling of the steam. The engine will thus run out at a moderate speed and it is at such moderate speeds that the throttling action is most beneficial. This throttling action will be similar to that of a governor. It will increase the diagram, when the speed decreases with high loads. On the other hand a decreased load will increase the speed and the diagram will also be decreasad. The throttling action is quite adequate to prevent the engine reaching excessive speeds. The engine is stopped by setting the valve gear over to the middle position.

As mentioned above, the valve gear fulfils the above requirements by using an auxiliary inlet with a late cut-off and a main inlet with an early cut-off. In the present instance, (fig. 193), a Walchaert valve gear has been chosen because it lends itself very well to the purpose, as it is possible with this gear to give different angles of advance and cut-offs for each set of valves. The valve rods are coupled to different points of the Walchaert lever so as to give the desired differences of cut-off.

Figure 194 shows a form of reversing rolling mill engine in which the motion of the gear is transmitted to a rock shaft arranged in the centre of the cylinder and from this rock shaft to piston valves arranged in the cylinder covers. In this case also, auxiliary valves are employed to give the above mentioned results, which correspond to the requirements in practice.

Figure 195 shows a reversing una-flow rolling mill engine built by the well known firm of rolling mill engineers, Messrs. Ehrhardt & Sehmer. In the construction shown, the makers retain their well tested design in all its details, whilst employing the una-flow engine cylinder.

A valve operating shaft, supported by the frame, is driven from the main shaft through tooth gearing. On this valve shaft, eccentrics for a Stevenson link gear with crossed rods are mounted. These eccentrics operate piston valves which control the inlet. The exhaust is controlled by the piston over-running the ports. In this engine the water hammering present in the ordinary engine of their former model is completely avoided. A separate port is provided to work with a different lap on the valve which is designed to give a later cut-off with an earlier admission. In this special port, which leads to the cylinder, a throttling valve is introduced, which is regulated so as to give the proper amount of throttling.

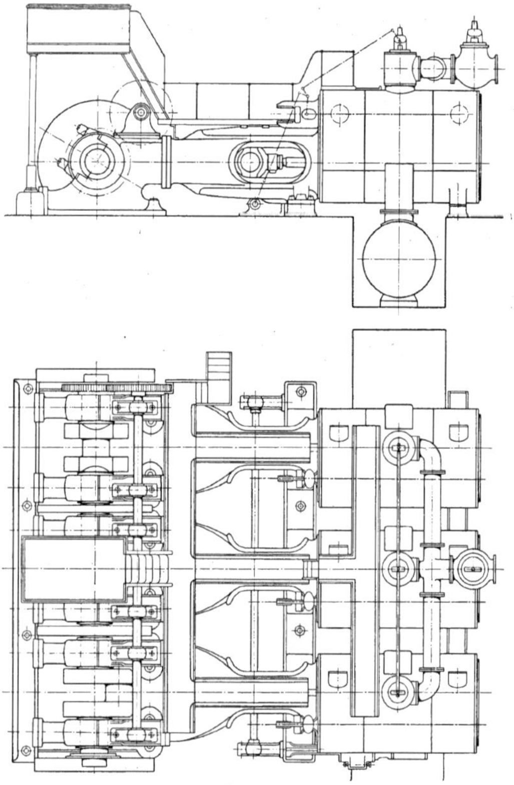

Fig. 195.

Messrs. Ehrhardt & Sehmer used to build these rolling mill engines as six cylinder engines arranged in tandem pairs. By employing the una-flow cylinder, three cylinders may be dispensed with, as well as three intermediate pieces and receivers with their accessories.

The valve control is effected from a single lever which adjusts both the inlet valve and the link in such a manner that when the link is in the middle position, the inlet valve is closed, and when the link is at the ends, the valve is given a full opening. When the link is in its mid-position, the lever connections from the bottom shaft to the inlet valve are stretched to their furthermost extent. The link is set from this same bottom shaft. The engine always works with the full steam pressure in the valve chest. The power exercised by the engine is controlled by regulating the cut-off. The cut-off diagrams correspond to the requirements of normal rolling and this method results in a considerable saving of steam. The throttling mentioned above, as required for starting and stopping the engine, is given by the arrangement of the above described auxiliary inlet port which works with a different lap to the main port. The later cut-offs obtained by this auxiliary gear, which is operated simultaneously with the main gear, produce throttling at starting and stopping, so that excessive speed and excessive steam consumption are effectively avoided. By employing a throttling valve in this auxiliary port, it is possible to adjust the throttling so as to arrive exactly at the best conditions for working. In the designing of such auxiliary gears, the designer has to take care that the admission is not too early, because, if the admission takes place too early before the end of the stroke, the engine does not work smoothly.

An examination of the diagrams, as well as experience gained from other una-flow steam engines, shows that engines of this kind have a very uniform turning moment, which is naturally of the greatest importance for reversing rolling mill engines employing no flywheel. In addition to this the simplicity of the una-flow cylinder will be appreciated as an important advantage. The constant compression and the adequate cushioning thereby obtained are of special value in the case of high velocities. In the ordinary rolling mill engine, the compression at early cut-offs is very considerable, and this compression makes the turning moment irregular and places an undue load on the driving mechanism and cylinder. Such excessive compression is especially present in the ordinary counter-flow engine with early cut-offs and crossed eccentric rods, where the virtual eccentric is small. This causes the exhaust opening to be insufficient, and the counter pressure in the cylinder to be excessive, especially when running with high speed.

As a result of the single stage expansion in the una-flow engine it is extremely powerful and, in fact, can pull through a bite where a compound engine would be brought to a stop. This circumstance renders the una-flow steam engine specially valuable for rolling mill work. All exhaust valve gearing is dispensed with, as the exhaust takes place through the piston-controlled ports. The ports are so large that the cylinder is fully exhausted even at high velocities. It avoids excessive counter pressures and thereby the loss and danger due to excessive stresses on the working parts and cylinder. As compared with the known form of tandem engines, the throttling, radiation and conduction losses between the two cylinders are avoided.

The una-flow rolling mill engine, of course, may be constructed as a twin cylinder engine, but in such a case, naturally, the disadvantages present in any twin engine as compared with the triple engine or three crank rolling mill are to be reckoned with.

Figure 196 shows a una-flow rolling mill engine with a flywheel.

Fig. 196.

Fig. 198.

This engine was built by Messrs. Ehrhardt & Sehmer for the Düsseldorfer Eisen- and Drahtindustrie where it is used for driving a set of wire rolls. The crank is overhung and the engine is coupled directly to the rolls. The inlet valves in the cover are operated by cam gear from a separate valve shaft. On this valve shaft the governor is arranged.

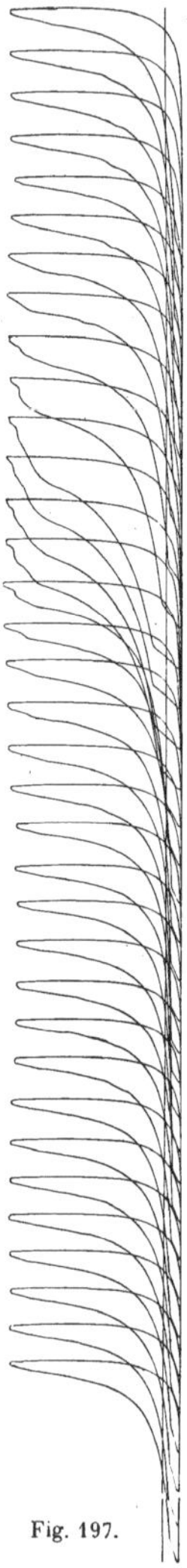

Fig. 197.

Vor dem Umbau.
Cylinderdurchmesser
Hub 1250 mm. Leistung
bei 60 Umdrehungen in der

Fig. 199.

Nach dem Umbau.
Gleichstrom-Walzenzugmaschine
Cylinderdurchmesser 950 mm,
Hub 1250 mm, Leistung 800-1150 PS.
bei 60 Umdrehungen in der Minute.

Fig. 200.

The diagrams taken during a "draw" (see figure 197) show that the steam distribution is excellent.

The governor controls the cut-off in the engine to meet the momentary conditions of the rapidly changing load on the rolls.

Messrs. Ehrhardt & Sehmer built a further una-flow rolling mill engine with a flywheel for the Gewerkschaft Quint of Trier. The cylinder is 630 mm diameter and the stroke is 1000 mm. The engine drives a set of preliminary rolls, a set of finishing rolls and also a set of polishing rolls (see figure 198).

The following results were obtained with this engine which proved quite satisfactory in practice:

Duration of trial	5 hours.
Revs. per minute	80.
Average I.H.P.	239 H.P.
Steam pressure in engine	9.5 atmos.
Steam temperature	191° C.
Temperature of the cooling water . .	11° C.
Vacuum	93%.
Steam consumpt. per hour	1330 kg.
Steam consumpt. per H.P. hour . .	5.57 kg.

The cylinder was not jacketted. The engine was built at a time, when the action of the steam jacket was not thoroughly tested. By using a steam jacket the steam consumption can be reduced by about three quarters of a kg.

The result obtained is extremely favourable especially in view of the fact that the jacketting is omitted.

Fig. 199 shows an old tandem rolling mill engine with flywheel. Fig. 200 shows the same engine rebuilt by Ehrhardt & Sehmer in Saarbrücken, the tandem cylinders being replaced by a una-flow cylinder.

A common practice was formerly to reduce the steam and coal consumption by adding a low pressure cylinder to single stage expansion engines. The same result may be effected by replacing an existing counter-flow compound engine by a single una-flow cylinder. The engine builder could, in such a case, claim as payment a part of the saving in coal effected by the change.

Another unaflow flywheel rolling mill engine was built by Ehrhardt & Sehmer for the Rombacher Hüttenwerke in Lothringen. This engine is remarkable on account of its size, the cylinder diameter being 1080 mm, and the stroke 1300 mm. The normal output is about 1500 HP. (in one cylinder), when running at 120 revs. p. M.

Chapter XII.

Una flow winding Engine.

1. Condensing.

A una-flow steam engine may be adapted for use as a winding engine by cutting out the excessive compression for a short time to facilitate the exact stoppage of the cage. This may be effected by employing a supplementary exhaust valve (see figure 201). The supplementary valve should have its exhaust lap so proportioned, that compression is almost nil during the time the cage is being brought into the exact position required, whereas, while the cage is travelling the full compression is in operation.

During travelling, therefore, the supplementary exhaust valve is out of action as regards its effect upon the compression. The exhaust is then controlled entirely by the ports which are covered and uncovered by the piston, so that the full compression (about 90%) of the una-flow engine is operative. According to the diagram in figure 202, the compression is about 25%, 65% and 90% of the stroke when the cut-off is 80%, 50% and 40% respectively. This shows how the una-flow engine may be arranged to work economically whilst still permitting of easy manoevering. When the main inlet valve is a slide valve, the exhaust lap may be arranged, in connection with the main inlet valve, to come into operation only when the cut-off is late, as described above. Figures 203 and 204 show a design for a small winding engine. Instead of the auxiliary exhaust slide valve shown in figure 201, two exhaust lift valves are provided at the ends of the cylinder to come into operation at late cut-offs. The advantage of this arrangement over that shown in figure 201, is that the clearance volume and surfaces are reduced. The valve gear is of the Gooch type and the motion is transmitted to both the inlet valves and the auxiliary exhaust valves, by means of a roller-cam gear. The valve diagrams for this engine are similar to those shown in figure 202. It is quite possible with an auxiliary exhaust valve with suitable lap, to work an engine of the above construction for a short time as a non-condensing engine. If it is the exception that the engine should work as a non-condensing engine, the exhaust valves may still be retained small, and for temporary purposes high steam velocities would be quite permissible.

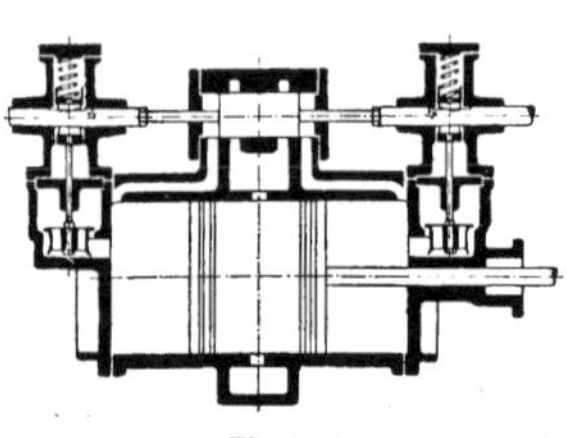

Fig. 201.

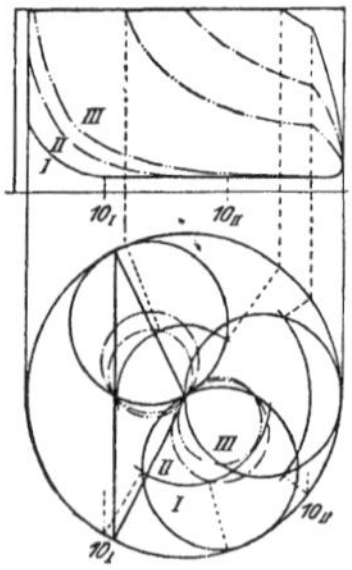

Fig. 202.

The force necessary to operate this gear when manoeuvring, is so small that stopping, reversing and adjustment may be effected by hand, thereby dispensing with a power manoeuvring engine. For this reason the Gooch link gear is advisable, as in this gear only the block and not the link itself has to be adjusted.

The four valves may also be operated by means of the ordinary conical cam gear (see figures 205 and 206). These conical cams are preferably so arranged that the inlet valve is only raised a short distance when the cut-off is at 9/10ths. of the stroke, and the auxiliary exhaust valve should remain open till very shortly before the end of the stroke. This enables the cage to be accurately adjusted. When the cage is travelling, the cut-off is earlier and the auxiliary exhaust valves are out of operation. These exhaust valves are only in effective operation when the cage is to be brought to a standstill at any desired point.

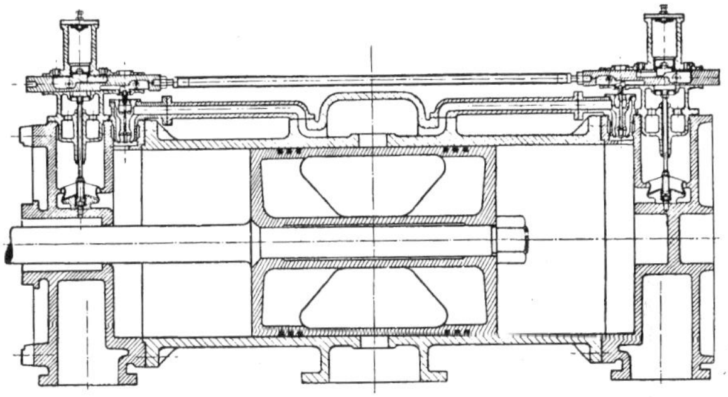

Fig. 203.

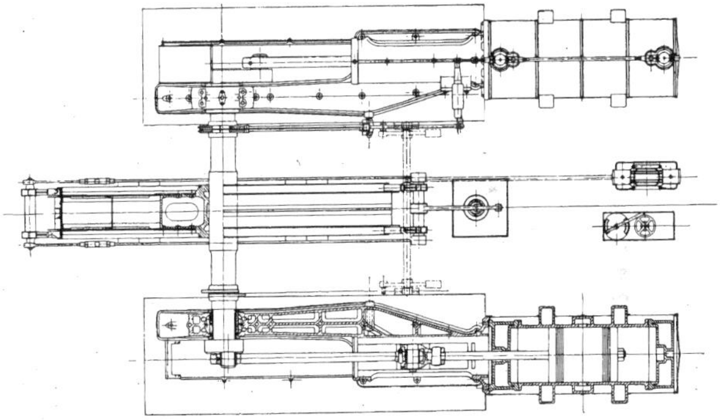

Fig. 204.

During ordinary travelling therefore, the engine works as a pure una-flow engine, and for the greater part of its working time, it operates under the most economical conditions.

The conical cam gear may also be operated without an auxiliary power cylinder. The clearance space may be calculated for compression extending over 9/10ths. of the stroke so as to give the proper degree of compression for all reasonable condenser pressures. The manoeuvring of the engine may be effected very freely and easily without employing safety or spring loaded valves for preventing excessive compression pressures.

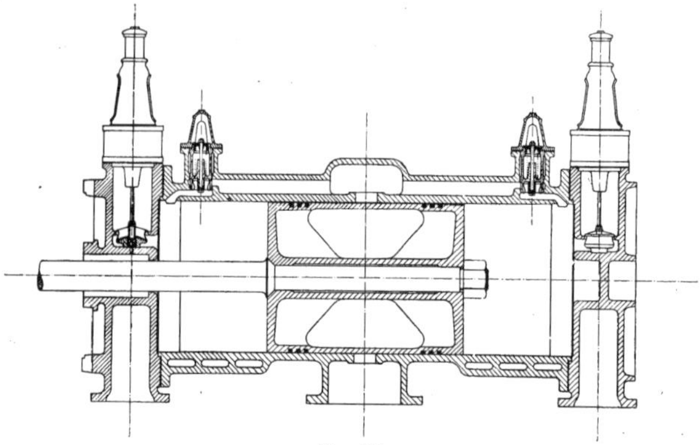

Fig. 205.

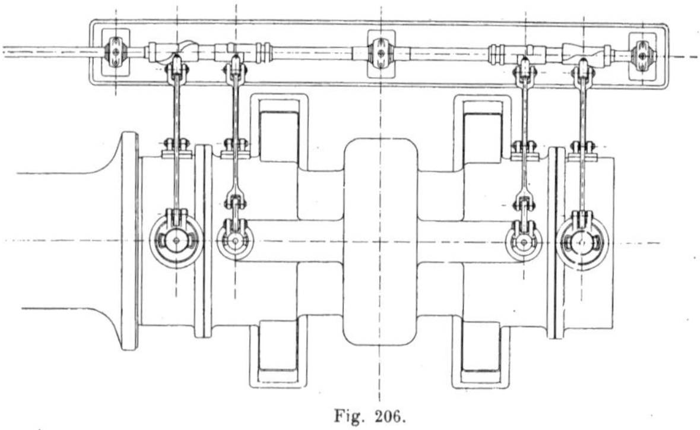

Fig. 206.

2. The una-flow winding engine working with atmospheric exhaust, or exhaust to a low pressure turbine.

When the una-flow winding engine is to work with atmospheric exhaust, or with exhaust to a low pressure turbine, the type of engine shown in figures 103 to 105 may be employed, that is to say, a una-flow engine with an auxiliary or supplementary exhaust valve in the piston. In this case the inlet valves as well as the exhaust valves in the piston must be reversible.

A una-flow winding engine with this arrangement is shown in figures 207 to 208. The inlet valves are operated in the ordinary manner by conical cam

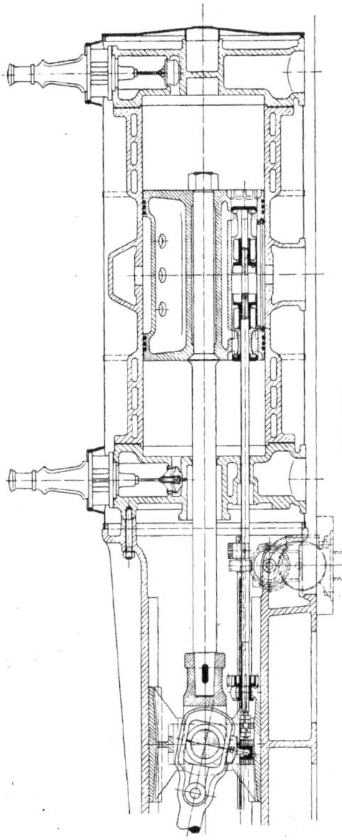

Fig. 207.

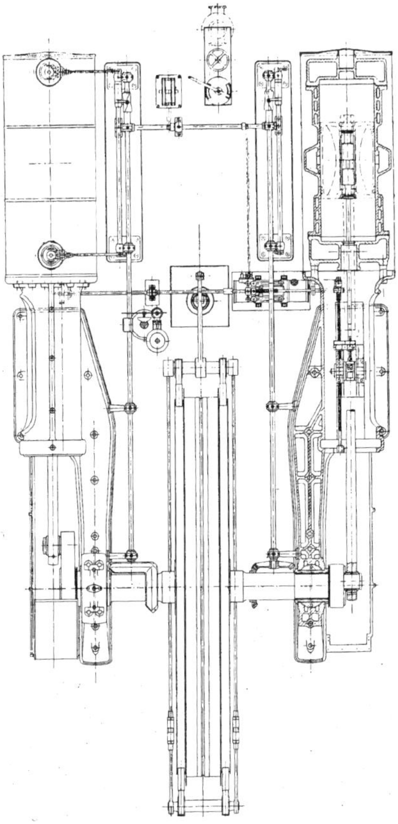

Fig. 208.

gearing. The exhaust valve is a piston valve, which, for the purpose of reversing, is divided into two parts. Each half of the piston valve engages with a screw thread on the rotatable valve spindle, the two screw threads being respectively right and left handed. When the spindle is turned, the cut-off edges of the valves are so adjusted that when the engine is running forwards, the inner edge of the valve co-operates with the outer controlling edge on the piston, and during reversing the outer edge of the valve co-operates with the inner controlling-edge on the piston. The rotation of the exhaust valve spindle is effected by means of a motor cylinder, which rotates a cross shaft, through quick pitched screw gear, and this cross shaft transmits its rotation to a counter shaft arranged parallel to the cross-head guide. The counter shaft operates the valve spindle through toothed gearing, located on the valve spindle and on the counter shaft. During manoeuvring, the adjusting gear for the valve cam is coupled to the above mentioned power cylinder so that both the inlet valves and supplementary exhaust valves in the piston are manoeuvred simultaneously. With this design, the exhaust valves always work in one or other of their extreme positions, and thereby always give their

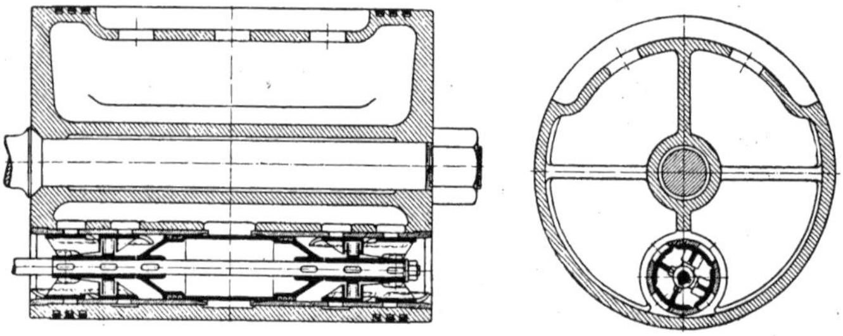

Fig. 209.

maximum port area for the auxiliary or supplementary exhaust. There is thus no throttling action introduced at intermediate points when manoeuvring. The pressure fluid cylinder may be replaced by an electrical controlling device. The inlet valves could be operated quite well by any ordinary type of link motion, and such link motion could be coupled up to the reversing gear for the exhaust valves. The exhaust valve may also be arranged as a Rider valve (see figure 209), but this involves increase in the clearances.

A winding engine, with a valve in the piston such as described, would always work as a pure uni-directional flow engine and would consequently always retain the thermal advantages peculiar to this type of engine.

Figure 210 shows a una-flow winding engine built by the firm Gutehoffnungshütte for the Vondern colliery pit No. 2. The engine is built to work eight cages, each with a working load of 500 kg. at a depth of 600 meters and a maximum velocity of 20 meters per second. It is provided with two rope drums of 6400 mm. diameter and 1900 mm. width. At present the engine is working with an admission pressure of 8 atmos. and is a non-condensing engine. The non-jacketted cylinders are both 1100 mm. diameter and 1600 mm. stroke and are supported at their ends on feet set on the bed plate. The inlet is controlled by piston valves which, during adjustment of the cage to any desired position, and during starting, also act as supplementary exhaust valves. These piston valves therefore perform the function of relieving compression to enable exact stoppage of the cage.

In this engine there is no exhaust valve in the piston, although the engine works with exhaust to the atmosphere. The clearance space is 10% and by employing an auxiliary clearance chamber it can be raised to 18%. The valves are moved through a short cross shaft arranged about the middle of the cylinder and

Fig. 210.

driven from the main shaft through a transmission shaft and gearing. This short cross shaft carries the conical valve cams. The valve cams control auxiliary valves on both sides, and these auxiliary valves in turn control the main piston valves through a servomotor connection. The conical cams are adjusted by hand directly from the controlling lever, without any power mechanism. If the inlet valve in this case were arranged in the cover, a substantial improvement would result. Although the conditions in the case just described were very unfavourable for obtaining a good tangential effort diagram, the engine nevertheless gave excellent results in this respect. In this connection, the ample inertia of the rope drums played a not unimportant part. Even when the rope speed was low, as is necessary when inspecting the rope, the engine performed its duties satisfactorily.

In practice, this engine proved that the cut-off could be made early, without in any way damaging the rope. This fact is of importance for condensing engines with which a specially uniform tangential effort diagram is obtained.

The author would refer the reader to the flywheel investigations given in Chapter VI, which show that in condensing una-flow engines, the flywheel weight is much lower than in ordinary tandem engines, and with ample flywheel weight, more silent and even running of the rope would naturally be the result.

The winding engine is provided with a depth indicator, which acts as a safety apparatus, and is of the type now generally used by the Gutehoffnungshütte. The depth indicator governs the winding pull from the beginning to the end of the lift or descent, and ensures that the maximum permissible velocity is not exceeded. It further ensures that the velocity decreases at a predetermined rate before the cage comes to the pit head. The cage always travels to the pit head with a very low velocity. The safety apparatus controls the engine, through the steam brake, which acts with an increasing force proportional to the excess speed, and when the speed decreases the brake is correspondingly released. Shortly after the cage reaches the pit head, the steam brake acts with full force. This latter control is independant of the cut-off control during the initial stages of travelling. This ensures safety both when raising loads and during descent. Trials with this winding engine have shown that the steam necessary for adjusting the cage in relation to the steam required to raise the cage, is, even under the most unfavourable conditions, only 2%, a figure which is sufficiently insignificant.

The method adopted by the Gutehoffnungshütte (viz: to omit all valves in the piston for the sake of simplicity and to allow the exhaust to take place during normal running only through the piston-controlled ports), is capable of development in two directions. In the first place the exhaust pressure may be diminished, which is quite possible even when the exhaust steam is led to a steam turbine, and secondly, the boiler pressure may be increased, a step which is, with modern boilers, quite easy. In both ways the clearance space may be diminished, and the steam expansion increased. In the following table, the clearance spaces are given for intermediate pressures from 0.6 to 1 atmos. absolute, and initial pressures of 7 to 16 atmos. above atmospheric pressure.

Intermediate pressure in atmos. absolute	Initial pressure in atmos. above atmosphere									
	7	8	9	10	11	12	13	14	15	16
1	17.2	15.2	13.6	12.4	11.4	10.5	9.8	9.2	8.6	8.0
0.9	15.4	13.6	12.3	11.1	10.3	9.5	8.8	8.3	7.7	7.2
0.8	13.7	12.1	10.9	10.0	9.1	8.5	7.9	7.4	6.9	6.5
0.7	11.9	10.6	9.5	8.7	8.0	7.5	6.9	6.5	6.1	5.8
0.6	10.2	9.1	8.2	7.5	6.9	6.4	6.0	5.6	5.3	5.0

From the above table it will be seen how the clearances may be materially reduced by lowering the intermediate pressure, when a steam turbine is connected up to the engine, and also by increasing the boiler pressure. The basis for the design of a una-flow engine, working with counter pressure on the exhaust, can in this way be easily modified so as to suit those requirements which are, at present, deemed necessary for condensing engines.

The reduction of the steam pressure for the exhaust steam turbine would be quite possible if corresponding care were taken in the design of the condenser into which the turbine exhausts. Reference might here be made to the analagous case in marine practice, in which exhaust steam turbines working in combination with reciprocating engines have been used with a pressure in the intermediate receiver of about one half atmos. The introduction and adaptation of the una-flow principle in the design of winding engines has taken much longer time than the adaptation of this type of engine to other purposes. This is easily explained when the size of the engines is taken into account and the consequent great risk. The enterprise of the Gutehoffnunghütte deserves all the more recognition. The success obtained by this firm is a proof that they well deserve the high position they hold as builders and designers.

3. Comparison of the una-flow winding engine with other modern constructions.

The una-flow steam engine, generally speaking, has a steam consumption equal to a good compound or triple expansion engine. With a high superheat, a steam consumption has been obtained as low as 3.8 kg per I.H.P. The tests all go to show that the steam consumption in the case of the una-flow engine is not so much dependent upon the cut-off as is the case with ordinary multiple expansion engines. This fact leads one to expect great things for the future of the una-flow winding engine.

In the first place the una-flow steam engine is specially suited for the work of a winding engine on account of the direct action of the steam. The greater the number of stages in an engine, the more is the engine crippled, and conversely, the smaller the number of stages, the more energetic is the action of the engine. For this reason, and also on account of the greater completeness of the diagram the volume of the working cylinder space can be taken considerably less than that of a compound engine. This is felt not only when adjusting the position of the cage and during the time the cage is being accelerated, but also during ordinary travelling. The una-flow steam engine can work economically with a much higher mean effective pressure, than is possible in multiple expansion engines of the ordinary counter-flow type.

There is no need to worry over the difficulty of maintaining the compound effect when the load is varying, and when the engine is temporarily at rest for longer or shorter periods.

A considerable part of the radiation losses are entirely avoided in the case of the una-flow engine. The radiation losses are specially marked in the case of the four cylinders and the intermediate receivers used in a twin compound tandem winding engine. These losses are especially high in the case of such twin compound engines, where the pressure in the receiver is to be maintained during the longer or shorter periods in which the engine is at rest. Although the una-flow winding engine may require more steam for manoeuvring, this is more than counterbalanced by the greater radiation losses which occur in a compound winding engine of ordinary construction during the temporary periods of rest, and also by the trouble involved in relieving the rear side of the pistons of such compound engines from high counter pressures. A great part of the area of the diagram in the case of a compound engine is lost by throttling, and this has a serious effect upon the completeness of the diagram, and the efficiency of the engine. Taking everything into account, it may be stated as an absolute fact

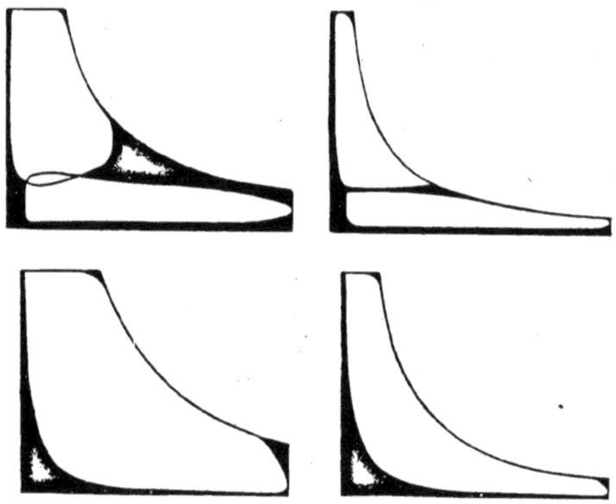

Fig. 211.

that, in view of the results obtained with protracted tests, the saving in steam is at least 15% in comparison with a twin compound winding engine of ordinary construction with the cylinders arranged in tandem.

Figure 211 shows comparative diagrams taken from a twin compound engine with tandem cylinders (900 mm. and 1400 mm. by 1800 mm.), and corresponding diagrams from a una-flow steam winding engine (1250 mm. by 1800 mm.). The larger diagrams represent those taken during the period in which the cage is being accelerated, and the smaller ones those taken during ordinary running of the cage. Judging these diagrams from a mere mechanical point of view, it will be seen that there is a decided saving on the part of the una-flow engine. The large cross sectional area of the exhaust ports prove specially advantageous, as they ensure a low counter pressure, even when the cut-off is very late, during the time that the cage is being accelerated.

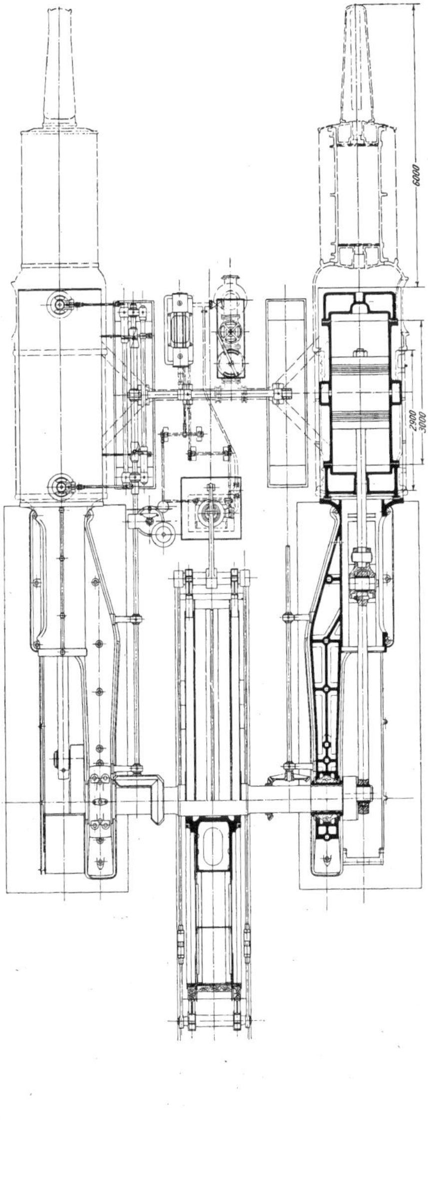

Fig. 212.

The efforts on all moving parts of the crank motion are of course larger in una-flow winding engines as compared with compound engines, as pointed out in a previons chapter.

In addition to the thermal advantages, the una-flow winding engine possesses also several prominent structural advantages. To show these clearly, a plan has been drawn in figure 212 showing the average decrease of over-all dimensions effected in changing a twin compound engine with tandem cylinders of 900 and 1400 mm./diameter respectively and 1800 mm. stroke into a una-flow winding engine of the same power. The una-flow winding engine would be 1250 mm./diameter and the length of the cylinder would be 3000 mm. as against 2900 mm. length for the low pressure cylinder of the compound engine. The over all length of the una-flow engine would be six metres less than that of the tandem engine and the engine house and the foundation would be reduced by about the same length. In addition two cylinders and their valve gears, two intermediate pieces and two receivers are dispensed with. The amount of lubricating oil used is correspondingly less, and the engine is considerably simpler, cheaper and surer in operation.

Chapter XIII.

The una-flow engine for driving compressors, blowers and pumps. Una-flow compressors and blowers.

The simplification of the engine due to avoidance of multiple expansion cylinders is a decided advantage when the engine is used for driving compressors,

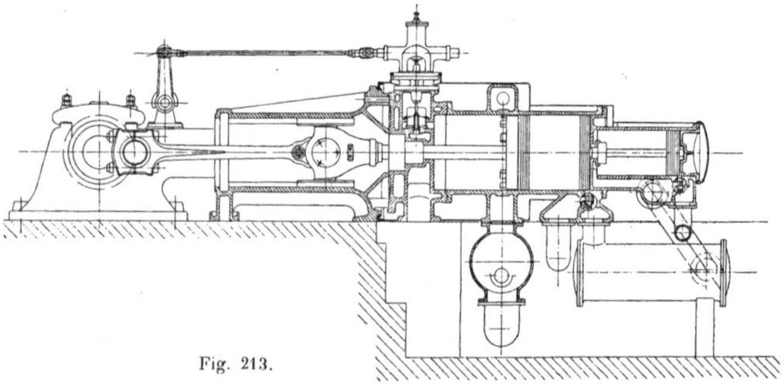

Fig. 213.

blowers or pumps. Figure 213 shows a single acting una-flow engine combined with a tandem compressor. The single acting low pressure cylinder of the compressor, and the single acting steam cylinder of the una-flow engine, are arranged in one.

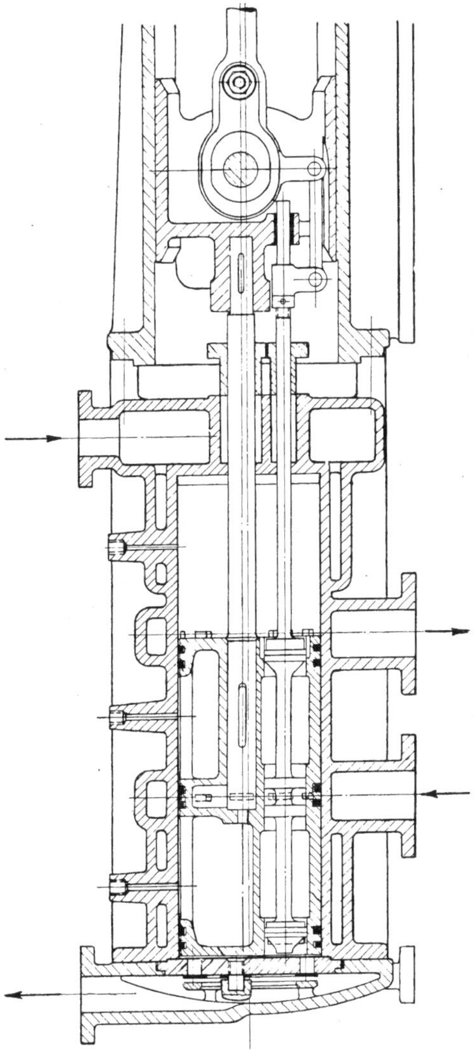

Fig. 214.

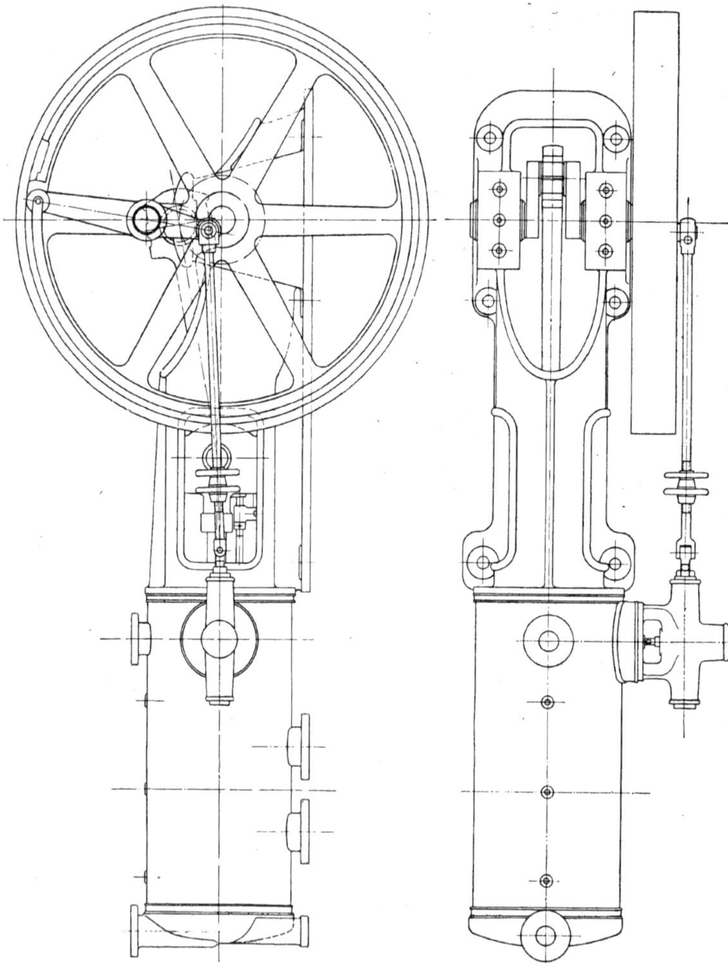

Fig. 215.

A piston rod connects the high pressure air piston to the combined steam and air piston. The inner side of the high pressure piston is in constant communication with the intermediate cooler between the stages of the compressor. A stuffing box is arranged in the gland, where the piston rod connecting the two pistons passes through the dividing wall between the two cylinders.

The second stage of the compression is effected on the rear side of the high pressure piston. During the suction stroke of the compressor, the atmospheric pressure on the rear side of the steam-air piston, counterbalances the compression on the front side of the steam-air piston, so that the work done by the atmosphere on the air side of the piston is approximately equal to the work done in compressing the steam on the steam side of the piston. During the air compression stroke, the energy of the steam overcomes the resistance on the two air pistons, so that during the air compression the opposite state of affairs exists to what existed during suction, namely, the energy in the steam counterbalances the work done on the air. This design is simple, cheap economical and certain in operation.

The una-flow air compressor, shown in figures 214 and 215, is specially interesting. In this case the single acting una-flow engine with supplementary or auxiliary exhaust is combined in one cylinder with a single acting una-flow compressor.

In the piston there is a slide valve of the construction described above, and this valve is driven from an arm radiating from the crosshead end of the connecting rod. The piston rod is eccentric, as the pressure on the surface of the piston valve in the piston is not transmitted through the piston rod. The principal reason, however, of this eccentric arrangement of the piston rod is to enable sufficient space for the insertion of the piston valve in the piston. This construction not only allows the two axes to be sufficiently separated from one another, but it also provides room for a long arm on the connecting rod end for operating the piston valve. After the controlling edge of the piston valve on the steam side has opened the exhaust ports, the piston valve allows exhaust to continue from the steam side until shortly before the end of the stroke. In the same manner, the piston valve on the rear side allows the air to pass into the air compression side as soon as the compressed air remaining in the clearance space from the previous stroke has expanded to atmospheric pressure. The admission of air is continued, after the piston valve is closed, by the piston uncovering the rear ports in the air end of the cylinder. Owing to the large cross section of these air inlet ports, the whole of the suction space is completely filled with air at atmospheric pressure. On the steam side, the port leading from the valve in the piston communicates with the exhaust ports, whilst on the air side the piston valve communicates with the air inlet ports. When the piston valve on the steam side opens the connection to the steam exhaust, the piston valve on the air side opens the connection to the air inlet.

The combined steam-air piston is provided with packing rings both at its ends and at the centre. The middle group of rings is for the purpose of preventing leakage between the steam exhaust space and the air suction space. The piston

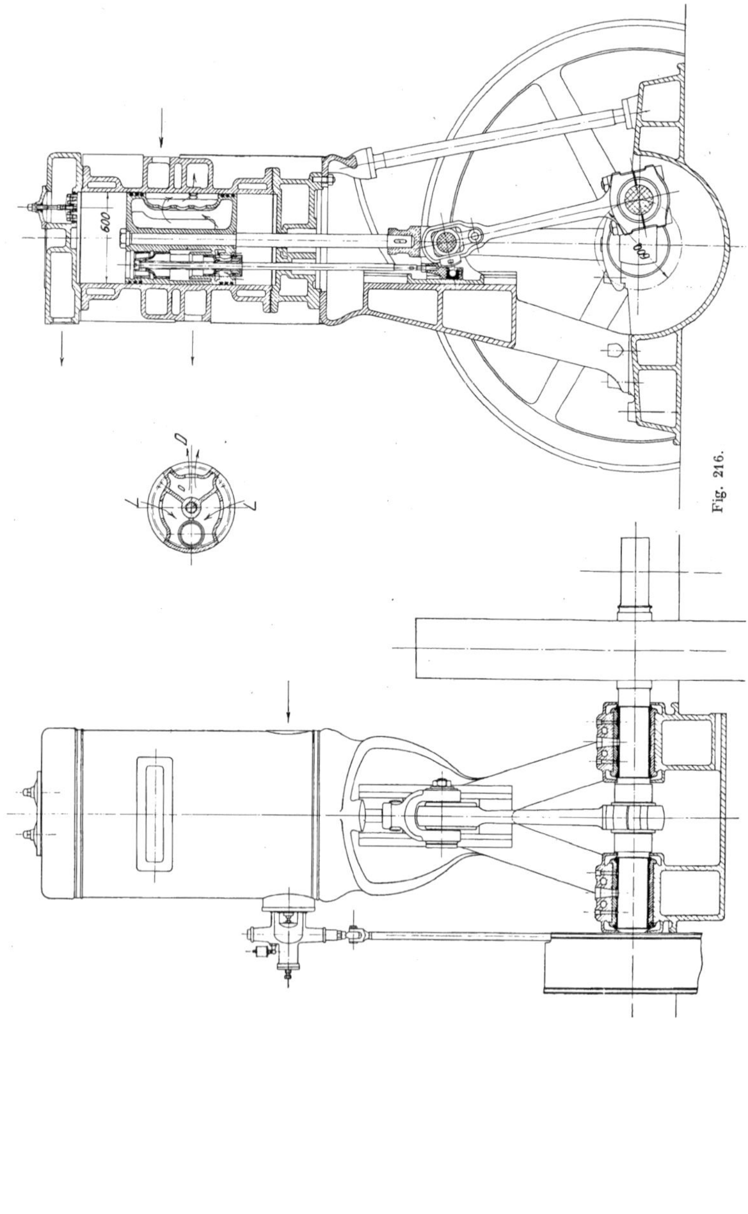

Fig. 216.

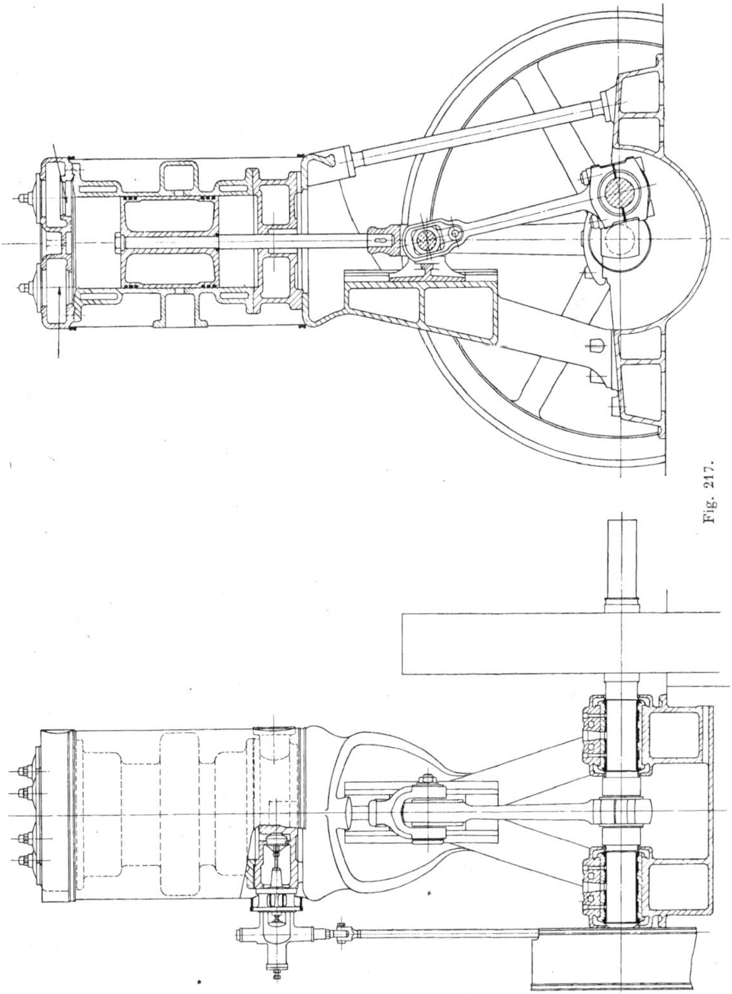

Fig. 217.

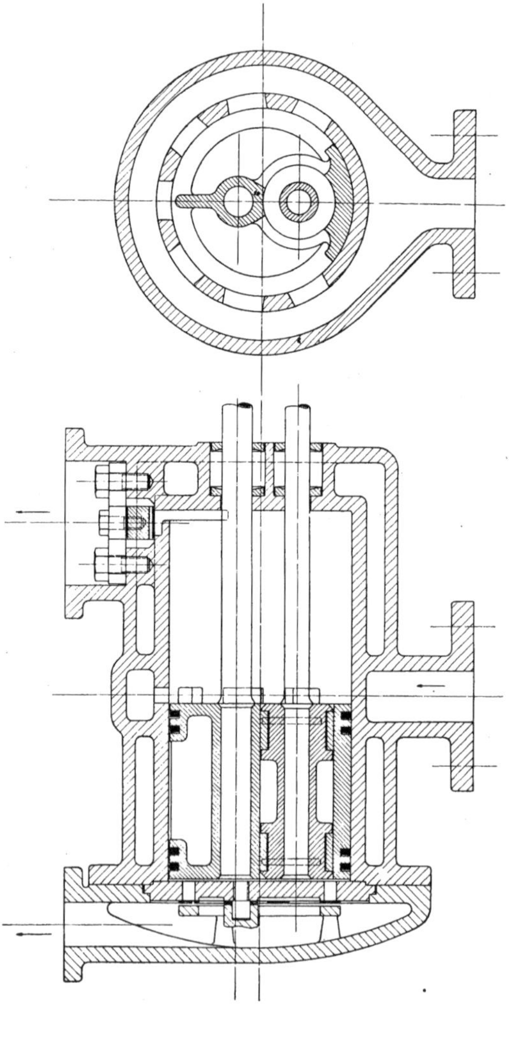

Fig. 218.

is double walled at this point in order to prevent as far as possible any heat transmission from the exhaust steam to the inflowing air. The pipe connections to the steam exhaust belt and to the air inlet belt are most conveniently arranged at the bottom in the foundation. This is especially so when it is desired to draw the air from outside.

It will be seen from the above description that the air compression side also works on the uni-directional flow principle, the air being caused to flow in the opposite direction to the steam.

The above described arrangement of inlet ports for the air, in which the incoming air flows through the hollow space in the piston and through the piston valve, is especially valuable in the case of small engines as it leaves ample space for the automatic discharge valves. The entire end of the cylinder remains free for this purpose as can be seen in figure 214.

The engine frame is of forked construction and the shaft is cranked, whilst the flywheel is overhung. A spring-controlled governor is arranged on the flywheel somewhat similar in construction to that described earlier with reference to figure 180. The governor controls an eccentric which operates the valve by means of a roller and cam gear. A hand wheel is arranged in the connection, so as to enable the engineman to adjust the length of the connection between the eccentric and the gear. Each compression stroke involves the same amount of work and consequently the same cut-off in the engine. When, therefore, the length of the valve rod is altered, the position of the eccentric must be changed, whereby the tension on the controlling spring of the governor is also changed. The changed tension on the spring causes a corresponding change in the centrifugal-force which results in a corresponding alteration in the number of revs. per minute and consequently in the output of the compressor. The designer has to arrange the stability of the governor spring so that the desired alteration of the output is obtained.

Figure 216, shows a vertical non-condensing steam-air compressor, in which both the steam and the air act on the uni-directional flow principle. In this case also, a combined steam outlet and air inlet valve is provided in the piston, this valve being operated by an arm on the end of the connecting rod. The air inlet ports, as well as the exhaust steam outlet ports, co-operate with the piston valve in the manner described above. The cross section of the piston is somewhat different as the air inlet ports are only arranged over part of the circumference of the cylinder, and the steam exhaust ports are arranged over the remaining part. In this case two longitudinal packings are provided on the piston so as to divide the steam outlet from the air inlet spaces. This division of the two spaces does not require to be so carefully effected as in the previous construction. Figure 217 shows a vertical compressor in which a single acting una-flow steam engine is combined with a single acting compressor of ordinary construction, that is with suction and compression valves. The steam end of the cylinder with its hot steam jacket is at the bottom, whilst the compressor with its water jacket is on top. The steam engine is condensing. If the compression on the steam side is sufficiently high, it is possible to balance the compression by the pressure of the inflowing air during suction, so that the engine runs fairly smoothly. In this case, as in the construction just described (fig. 216), the steam inlet valve is horizontal. The inlet valve is operated from an adjustable eccentric through a roller and cam gear.

A smaller una-flow air compressor is shown in figure 218, and this air compressor is driven by a belt or similar drive. The air in this case enters through inlet ports arranged in the middle of the cylinder. The air then passes to the interior of the piston, past the valve in the piston alternately to each end of the cylinder. After the air in the clearance space is expanded, the air enters past the valve in the piston, and the air admission is completed when the end of the piston uncovers the inlet ports. The large cross section opened in this way ensures the proper filling of the cylinder with air. The discharge valves may be arranged on the rear end of the cylinder which is entirely free, whilst on the front end they may be arranged to pierce through the cooling jacket at the side of the cylinder.

The piston valve is operated by an arm on the connecting rod end as described above. This combination of the piston-controlled ports with the piston valve control produces the retardation in opening and closing of the suction valves necessary to cause both the opening and closing to fall clear of the dead points.

A una-flow compressor of this kind has the great advantage that the incoming air is divided from the outgoing air, so that the air enters the cylinder at that part which is at lowest temperature. The heating of the air by the parts subjected to the heating action of the highly compressed air, is completely avoided. A further advantage resides in the excellent cooling action which is rendered possible by the additional length of the cylinder.

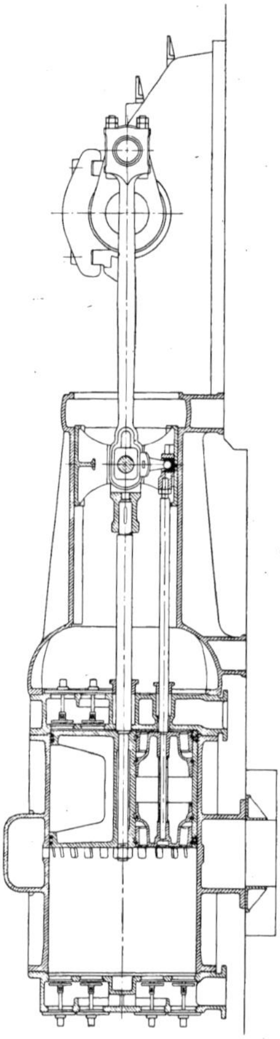

Fig. 219.

In figure 219, there is shown a una-flow steam engine arranged to drive a blower. The una-flow blower is constructed on the principles described above. The air entering through the suction ports arranged in the middle of the cylinder,

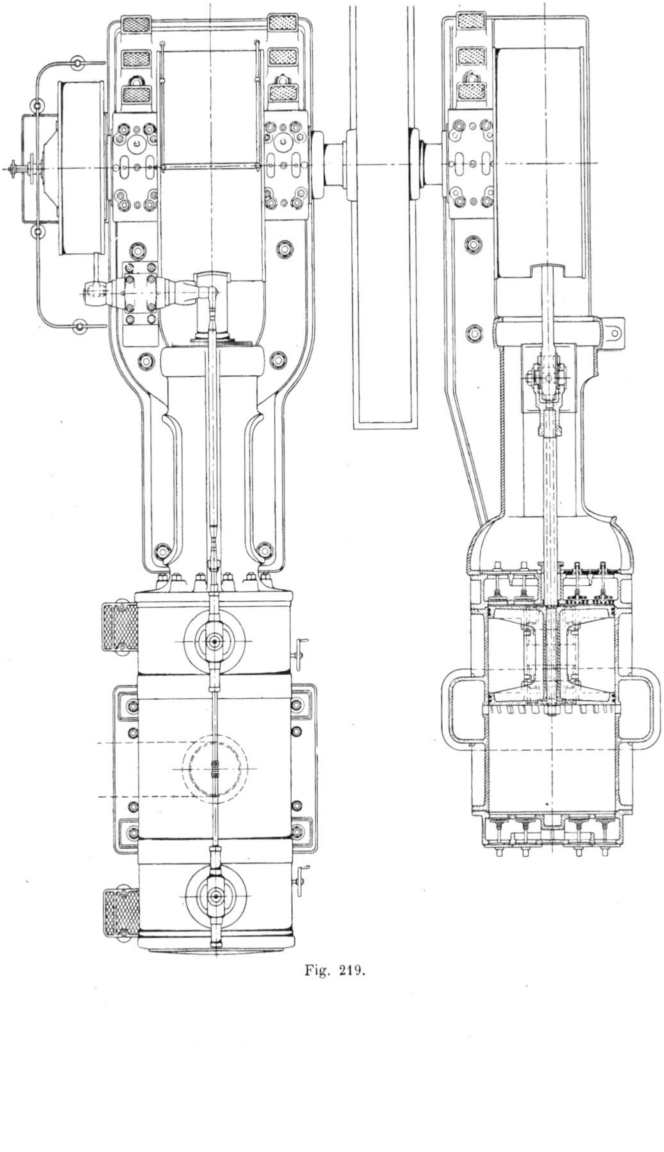

Fig. 219.

passes first through the piston and piston valve into the cylinder. The final filling of the cylinder is done through the ports in the cylinder walls uncovered by the piston. The discharge valves can be made of ample area, as the entire cylinder end is free for these valves. In consequence of this, the compressor may work at high speeds. In the case of a blower, special advantages accrue from,

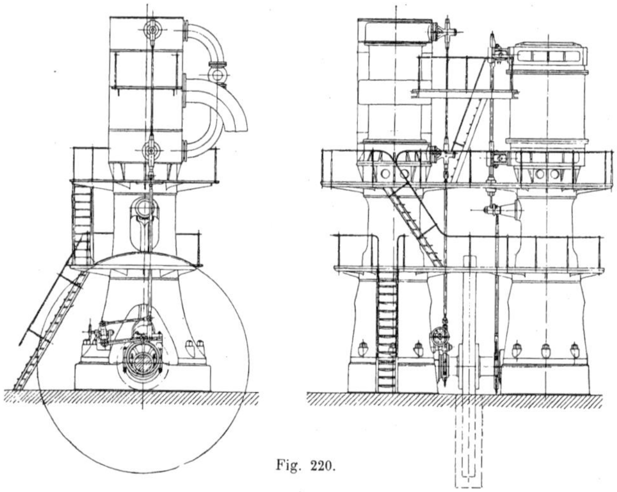

Fig. 220.

I. the complete filling of the cylinder due to the large cross section of the suction ports, II. the absence of all heating of the air during suction and III. the small resistance due to the large cross sectional area of the discharge valves.

Figure 220 shows a vertical una-flow engine with a vertical blower, the blower being of somewhat different construction.

In both of the above described blowers the cranks should be set so as to reduce the weight of the flywheel as much as possible.

Chapter XIV.

The una-flow engine for driving stamps and presses.

The una-flow steam engine with an exhaust valve in the piston may be very easily accommodated in the frame of a briquette pressing machine as can be seen in figure 221, plate 5.

In the manufacture of briquettes, a good drier is a necessity, to meet which exhaust steam is used with a pressure of 2 to 3 atmos. Accordingly, for bri-

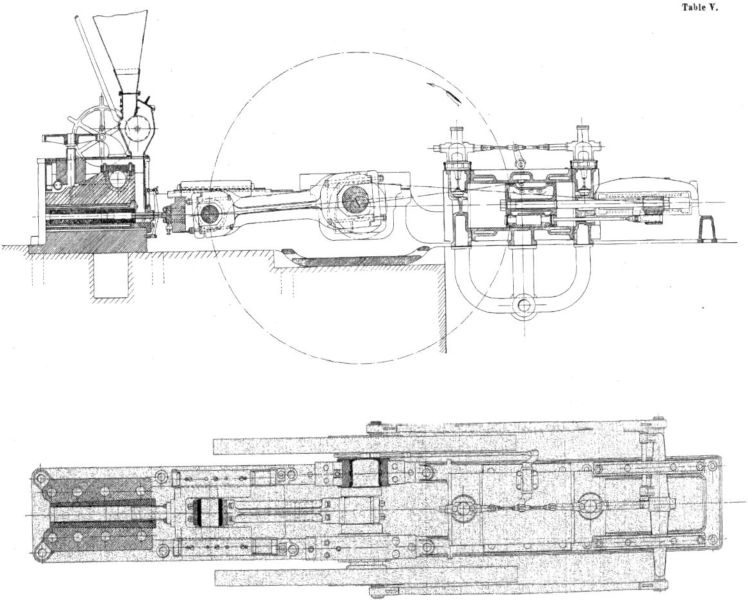

Fig. 221.

Printed by R. Oldenbourg, Munich and Berlin.

quette presses, it is necessary to have a una-flow steam engine with a prolonged exhaust. The entire absence of exhaust valves proper in the flat frame is a particular advantage, as such exhaust valves, together with the exhaust connection, would cause complications and would render the parts inaccessible. A short stroke, single acting briquette press, with its hopper, is arranged at the end of the engine. The stamp of the press is driven by a connecting rod from a crank on the main shaft. The main bearings are arranged one on each side of the crank, the flywheels being overhung. The crank pins are fixed to the overhung flywheels. The two connecting rods engage with a crosshead, which is guided horizontally and is connected at its centre to the piston by means of a piston rod. The engine cylinder is between the two flywheels near the main bearings whilst all the parts including the main bearings, the briquette press and the crosshead guides are carried on a single flat bed.

One of the flywheels carries a shaft governor for controlling the eccentric which operates both valves through a rocking lever supported over the cylinder. The exhaust valve in the piston is operated in the manner described above, by means of an arm radiating from one of the connecting rods and adapted to rock a countershaft supported in the crosshead. In plants of this description the steam used is usually saturated or only moderately superheated, and, as mentioned above, the una-flow steam engine is specially adapted for use with such steam.

Compound engines are too complicated for use in driving presses. Ordinary single stage expansion engines are undesirable on account of their inefficiency. The una-flow engine, however, unites in itself the simplicity of the single stage expansion engine and the efficiency of the compound engine and is in consequence specially suited for work of this class. To this has to be added other properties possessed by the una-flow steam engine in a high degree, such as the facility with which steam may be withdrawn for drying purposes. The una-flow cylinder being arranged between the flywheels, lends itself also to a compact design for the entire plant.

Chapter XV.

A una-flow marine engine.

In recent marine practice, serious endeavours have been made to introduce superheating and to employ balanced lift valves for distributing the steam. The una-flow engine is specially adapted to meet this modern tendency in marine engine practice, because it is well suited for working with superheated steam and also with balanced lift valves. It is also apparent that any advantage, which balanced lift valves may have over slide valves, is increased manifold when the exhaust valves are omitted, as is the case with the una-flow engine. The una-flow engine avoids that undesirable complication associated with the introduction of lift valve gear which hitherto has been the great stumbling block in the introduction of such gear in marine engines. The simplification introduced by the una-flow engine may be said to be of special advantage when the working medium is superheated steam. Owing to the unequal distribution of the superheat in the case of multi-stage engines many difficulties have arisen in practice chiefly in the high pressure cylinder. With a una-flow marine engine, on the other hand, there is little chance

Fig. 222.

for such difficulties, because the superheat is distributed equally on all cylinders and moreover the cycle in each cylinder is carried ont to a considerable extent with saturated and moist-steam. This increased reliability of the una-flow engine is of great importance for marine engines. The first una-flow marine engine was built because it was decided that this type of engine lent itself better to the introduction of superheating for marine purposes. It was thought that the una-flow steam engine solved this problem in the easiest and surest manner.

As mentioned in a previous chapter, the una-flow steam engine is also very well suited for working with saturated steam. For this reason, the new engine can quite well be introduced without violating the natural conservatisen of those ship owners and engineers who are still rather sceptical in regard to the intro-

Fig. 223.

duction of superheating. Such engineers always refer to the absolute necessity of excellent cylinder lubrication which is indispensable with high superheats and which endangers the safe and efficient working of the boiler. This is especially the case in multi-stage engines, where the most difficult working conditions are in the high pressure cylinder, which must be specially well lubricated. When saturated steam is used, cylinder lubrication may be entirely dispensed with, or the cylinder need only be lubricated at the beginning and end of a trip. Such a practice could really only be adopted, with safety, in a una-flow engine working with saturated or moderately superheated steam, as neither the cylinder nor the balanced lift valves require to be lubricated.

The first una-flow marine engine which has been built, is shown in fig. 222.

The Stettiner Maschinenbau-A.-G. Vulkan decided nest to try a una-flow marine steam engine for a steamer of their own build (see figures 223 to 228). The engine cylinders were 580 mm. diameter and 600 mm. stroke. With a pressure of 12 atmos. and running at 90 revs. per minute, the B.H.P. was 400.

Superheaters were provided and a mixing tube was also introduced so that saturated steam and superheated steam could be mixed to obtain a fairly wide range

Fig. 224.

of working superheats. The engine used had twin cylinders with cranks at 90°. The Klug valve gear was employed. The end pin of the Klug gear was connected by means of a bent rod to the vertical roller spindles for operating the valves which carried cams on their horizontal spindles. The Klug valve gear was designed for a maximum cut-off at a quarter of the stroke with sharp closure of the inlet valves. For the purpose of starting and manoeuvring, an auxiliary inlet slide valve was employed. It was driven from a second pin which in this case coincided with the pivot of the Klug eccentric rod. The auxiliary inlet valves were mounted

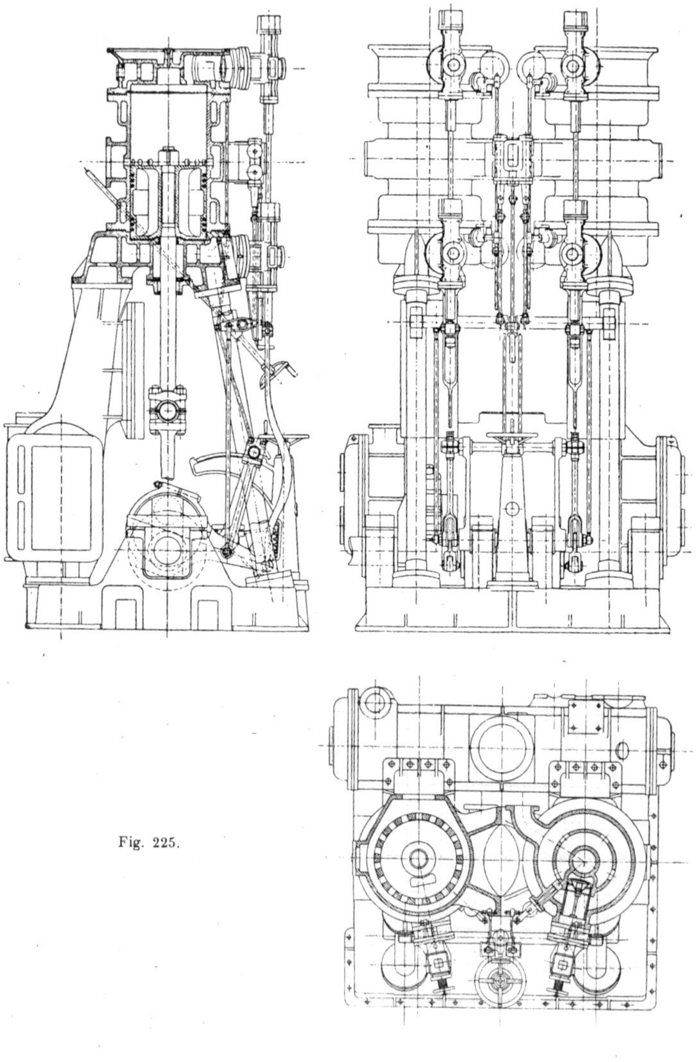

Fig. 225.

in a casing on the exhaust belt and they were arranged to work with late cut-offs up to about 90% of the stroke. The auxiliary piston valves also controlled an exhaust outlet, so that when starting without any vacuum in the condenser, excessive compression was avoided and starting facilitated. This type of auxiliary gear is shown diagrammtically in figure 226, which illustrates the arrangement employed on the unaflow marine engine of the »Strassburg« owned by the Hamburg American line.

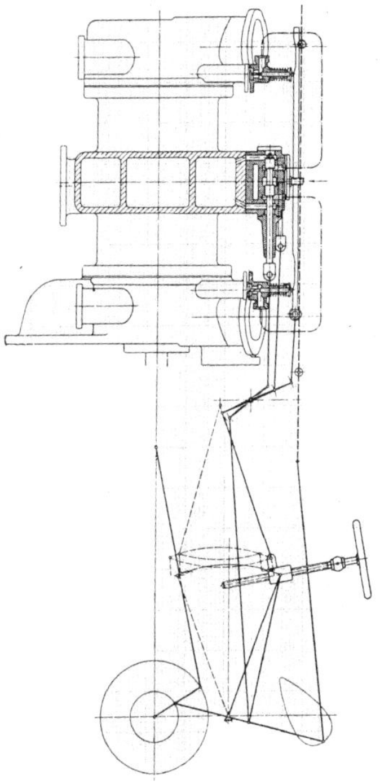

Fig. 226.

The automatic opening or switching in of the working steam connection to the auxiliary valve should be noted, as well as the valves for closing the connections from the ends of the main cylinder to the ports of the auxiliary valve. When the controlling lever of the gear is in either of its outermost positions, the switching valves for opening the inlet to the auxiliary piston valves and the connections to the ends of the cylinders, are opened.

In the intermediate positions of the lever these valves are closed and they remain closed until the lever approaches either of its end positions. If the early cut-off given by the main valve gear proves insufficient to meet the requirements of manoeuvring, the lever is set further over so as to bring the auxiliary gear more or less into effective operation.

After the engine is started, or the required manoeuvring is completed, the gear is set back until the position of normal cut-off, at about 10% of the stroke, is reached. When the lever is in this position, the auxiliary gear is completely cut out. There is no necessity for the engineer to operate by hand special auxiliary valves for manoeuvring, so that the operation and control of the engine is very much simplified, and no special difficulties can arise in emergencies. The condenser pumps are coupled to the main engine, and it is for this reason that the auxiliary inlet slide valve is also arranged to control an auxiliary exhaust so as to reduce compression when starting and manoeuvring. The condenser is in-

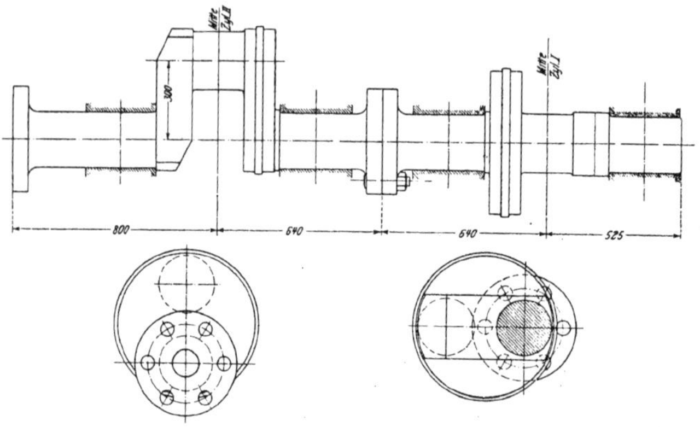

Fig. 227.

Fig. 228.

troduced in the ordinary manner into the rear columns. The entire valve gear is arranged on the front of the engine where all parts are accessible.

The eccentrics, as can be seen in figure 227, are arranged on the side cheeks of the crank. The engine is designed to work ordinarily with steam superheated to 250°. The cylindrical walls are accordingly provided with steam jackets near

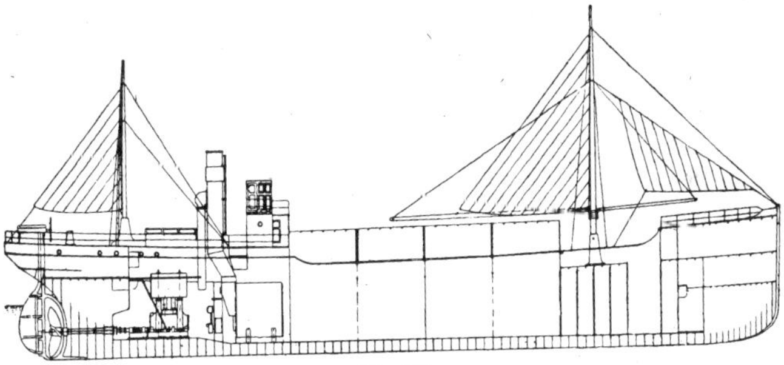

Fig. 229.

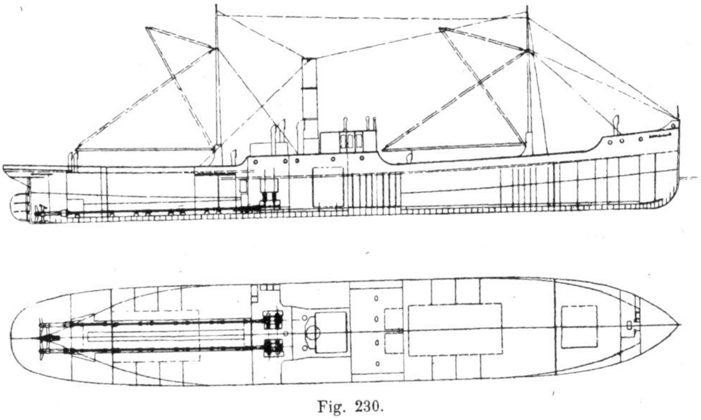

Fig. 230.

their ends, and of course the end covers are also provided with jackets. By connecting the cylinders together in the neighbourhood of the exhaust belts, where the temperature is lowest, the rigidity of the structure is increased, as it is not necessary at this cool point to make allowance for any serious expansion.

Figure 228, shows the valve bonnet with the valve.

Figure 229 is a side sectional diagrammatic view of the tramp steamer built by the Vulkan Co., for which the above described una-flow marine engine was built.

The Hamburg-America line decided to introduce two una-flow marine engines in the steamer "Strassburg", which plies between Hamburg and Cologne. This steamer was built in the yards of the Gebrüder Sachsenberg A.-G. Cologne Deutz

Fig. 231.

(figure 230). The ship has twin screws, each propeller shaft being driven by a vertical twin cylinder una-flow engine, the cylinders of which are 440 mm. in diameter and 450 mm. stroke. The steam pressure is 12 atmos, and the working temperature 325° C. The engine yields 250 I.H.P. at 175 Revs. These engines are built on the same lines as the Vulkan engines, as can be seen from the photographic picture in figure 231 and from the drawings given in figure 232. The engines use a Klug valve gear and auxiliary inlet as above described.

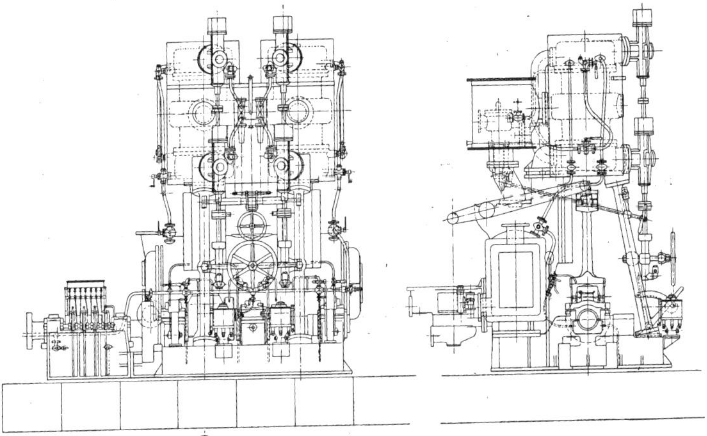

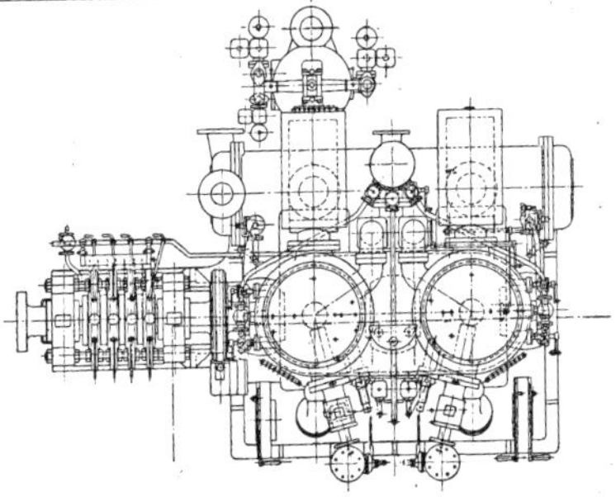

Fig. 232.

The auxiliary valve is operated from a different point on the eccentric rod to the main valves. On account of the high temperature of the superheat, no heater or jacket was arranged around the cylinder walls, but there was the usual heater on the cover.

Messrs. Burmeister & Wain of Copenhagen also decided to introduce the una-flow marine engine on two steamers ordered by the United Steamships Co. of Copenhagen. Each ship was fitted with a una-flow engine of 1000 B.H.P. These engines each had three cylinders, with Klug valve gear. In a three cylinder engine, the auxiliary inlet gear may be dispensed with, because the cranks being at 120°, the maximum cut-off necessary for manoeuvring purposes is only about 35% of the stroke. With a maximum cut-off at 35%, the inlet valves are closed sufficiently sharply at the normal cut-off of 10% of the stroke. As each cylinder is connected separately to the boiler on the one hand and the condenser on the other, it is possible

Fig. 233.

Fig. 234.

to cut out any individual engine. By using a later cut-off in the two remaining engines the full running power may be obtained.

Figures 233 and 234 show photographs of these engines. The compact arrangement of the Klug gear can be seen from these pictures. The operating rods are all arranged at one end of the engine and these rods rock three telescoping shafts which transmit the power to the various cylinders. The cylinders are 635 mm. diameter and 915 mm. stroke. The engine runs at 84 revs. per minute. The construction of the cylinders is such as to permit of them being arranged very closely side-by-side. The saving in length of the engine is about 1¾ meters. One advantage of this arrangement resides in the balancing of the weights of the reciprocating parts of the three engines. The cut-offs at the upper and lower ends of each cylinder may therefore be taken in all cases as equal. This results in very smooth running. The condenser pumps are coupled to the main engine. The vacuum for starting is created by a small auxiliary pump. Starting may also be facilitated by using an ejector or by opening the blow-off cocks. All the cylinders are provided with steam withdrawal valves, the withdrawn steam being used for heating the feed-water.

The connection between the exhaust belts of the cylinders and the condenser are specially worthy of notice. A separate connection is provided for each engine cylinder. As however, the exhaust belts are all interconnected there is always ample area of passage from each cylinder to the condenser. In spite of the high superheat used, there is no special provision for lubricating the cylinder. This is due to the fact that with a single stage expansion, even with high superheats, the working parts move over comparatively cold surfaces which are exposed for the greater part to saturated or moist steam. This results in a lower working temperature for the contacting surfaces. The engines are lubricated only very little during the whole trip or only at the commencement and at the end of the trip, whilst in ordinary running they are not lubricated at all. For this reason it was not deemed necessary to insert an oil separator between the engine and condenser.

One test, which was not merely an exhibition test, but a proper test under working conditions, showed a coal consumption of 0·6 kg. per I.H.P. hour. The coal was Newcastle coal with a calorific value of 7300 calories. The above figure included the steam used in the auxiliary engines. The steam pressure was 11 atmos and the temperature was 220. It should be noted also that there was no Howden force draft or heating of the air, and the walls at the cylinder ends were not jacketted. The covers only were heated. The superheaters were designed by Mr. Jörgensen and proved excellent in operation.

The ship proved most satisfactory both for the purchasers and the builders as the guaranteed speed was obtained with 800 HP instead of 1000 HP as mentioned in the specification. This was a pleasant suprise to both parties. The indicator diagrams taken during the trial trip are shown in figure 235. These diagrams show the satisfactory working of the valves.

Messrs. Burmeister & Wain may well be satisfied with the above, their first attempt. By using a steam jacket on the cylindrical walls, the coal consumption might well be reduced from 0·6 to 0·5 kg. per I.H.P.

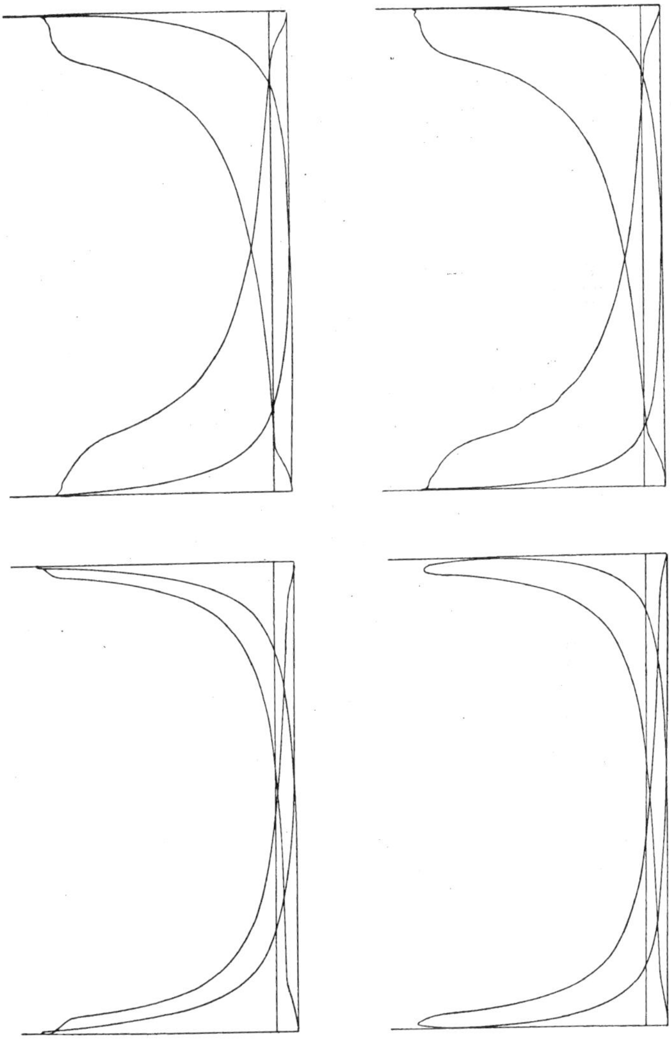

Fig. 235.

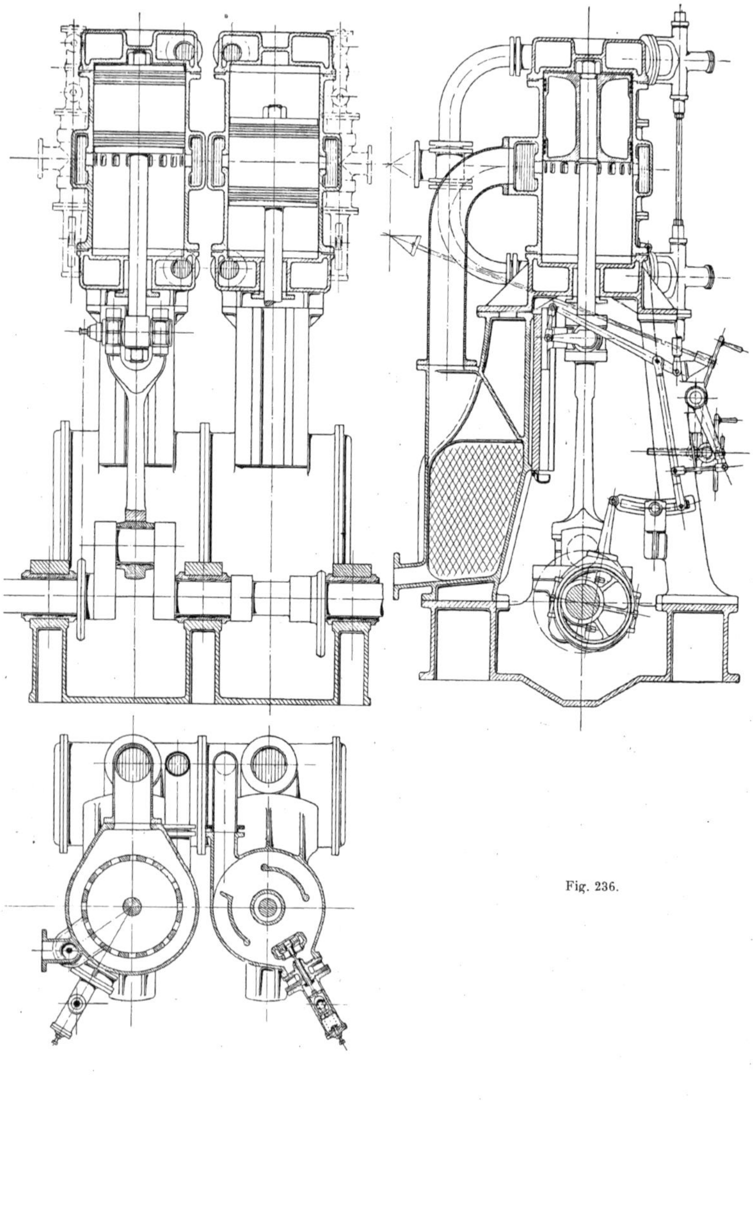

Fig. 236.

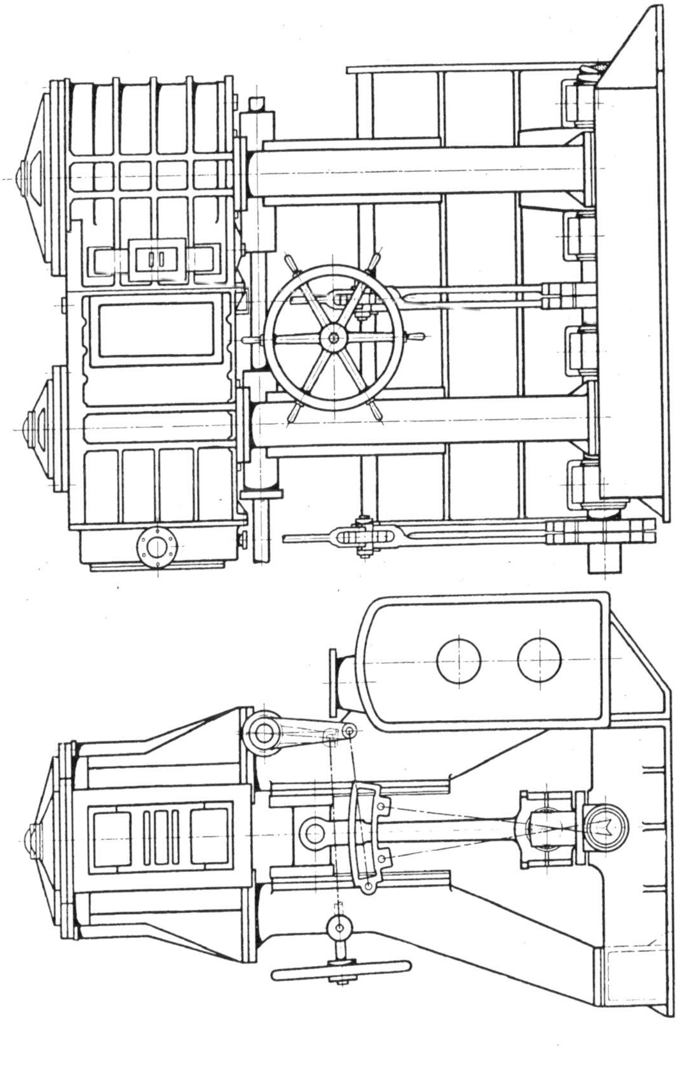

Fig. 237.

Figure 236 shows a proposal for a una-flow marine engine in which the Walchaert valve gear, which was found so successful in locomotives, is employed.

A compound engine of the Steamship "Wera" of the Orient Co., in St. Petersburg, is shown in figure 237. On account of the excellent steam consumption results obtained by the una-flow steam engine, this company decided to replace the compound engine by two parallel una-flow cylinders of 600 mm. diameter and 711 mm. stroke. Superheating had been tried experimentally on this ship but the old engine proved unsuitable for use with superheated steam. For this reason it was decided to introduce a una-flow steam cylinder of the construction shown

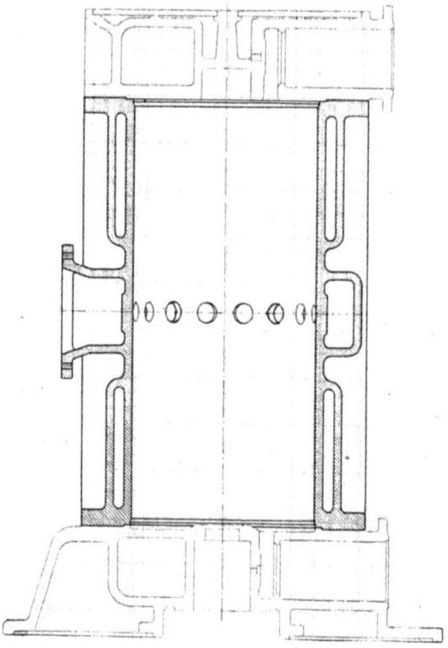

Fig. 238.

in figure 238 and 239. Two engines are provided in this steamer, each of which has to yield 500 HP at a maximum speed of 110 revs. per minute. The cylindrical walls are provided with steam jackets so that when desired the engines may be ready to work under the best conditions with saturated steam.

The valve gear shown in figure 240, is of special interest. A bevel toothed wheel is keyed on the free end of the shaft and drives another bevel wheel through the medium of a reversing transmission gear in a rotatable casing. The last mentioned wheel has fixed to it four eccentrics. Each of these four eccentrics operates one of the valves in the two cylinders.

The motion is transmitted through a rocking lever supported eccentrically and connected by a long rod to the rocking arms mounted on a bracket on the exhaust belt. The eccentrically supported rocking lever is coupled to the rotatable casing of the reversing gear so as to be adjusted simultaneously therewith.

The ratio of the adjusting mechanism for the rotatable casing and the eccentric pivot for the oscillating lever, is such that the casing moves through half the angle of the eccentric pivot of the oscillating lever. The operator adjusts the casing of the reversing gear and this adjustment is transmitted to the eccentric pivot of the oscillating lever through toothed gearing, thereby doubling the motion. Both the main eccentrics and the adjustable eccentric for the oscillating lever are moved in the same direction and in such relation that the proper lap is given for each

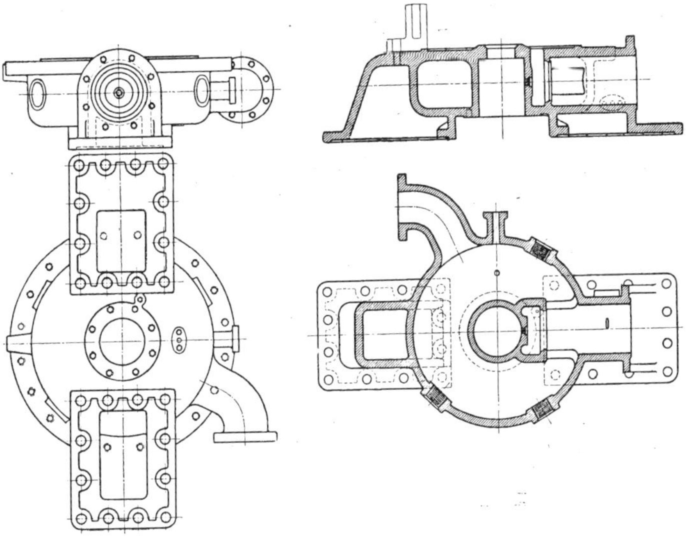

Fig. 239.

angle of advance. When the main eccentrics are switched over from the minimum forward running angle to the maximum reversing angle, the same adjustment takes place on the pivot of the oscillating lever. The curve for this gear is therefore not a straight line but a circle drawn about the centre of the shaft. The movement of the valve gear is always the same whilst the outside lap is altered in proportion to the angle of advance. When the oscillating lever has its arm in the ratio 1 : 1, the eccentricity of the pivot for this lever must be half the eccentricity of the main eccentric. The eccentric rods of the main eccentric are very short in order to compensate for the effect of the finite connecting rod. In addition, the oscillating levers are proportioned so as to give a later cut-off

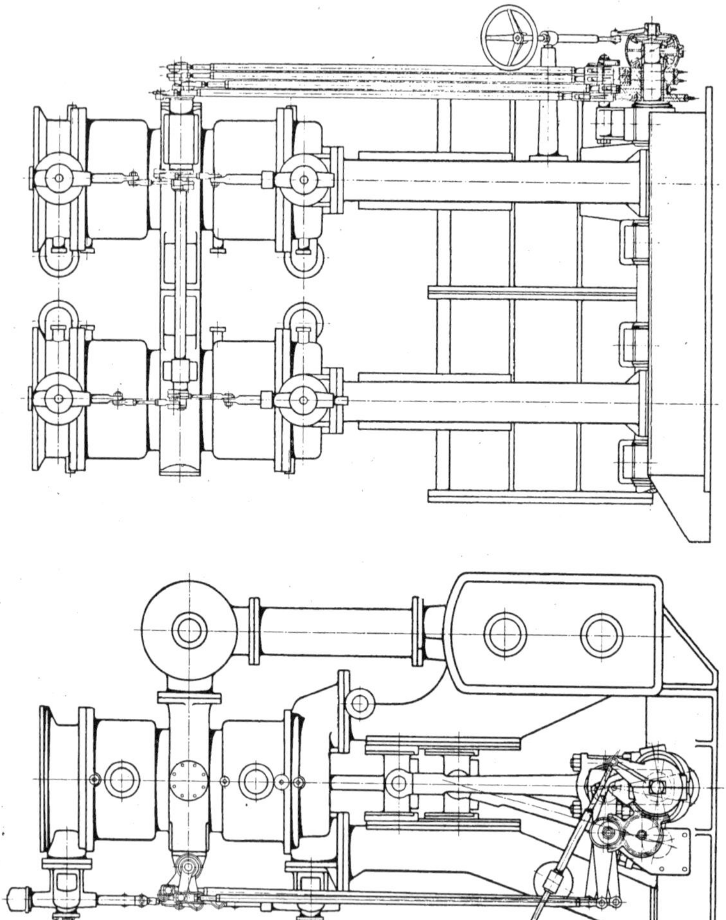

Fig. 240.

below and an earlier cut-off above during forward running. In this way the influence of the finite connecting rod and the weight of the reciprocating masses are compensated for.

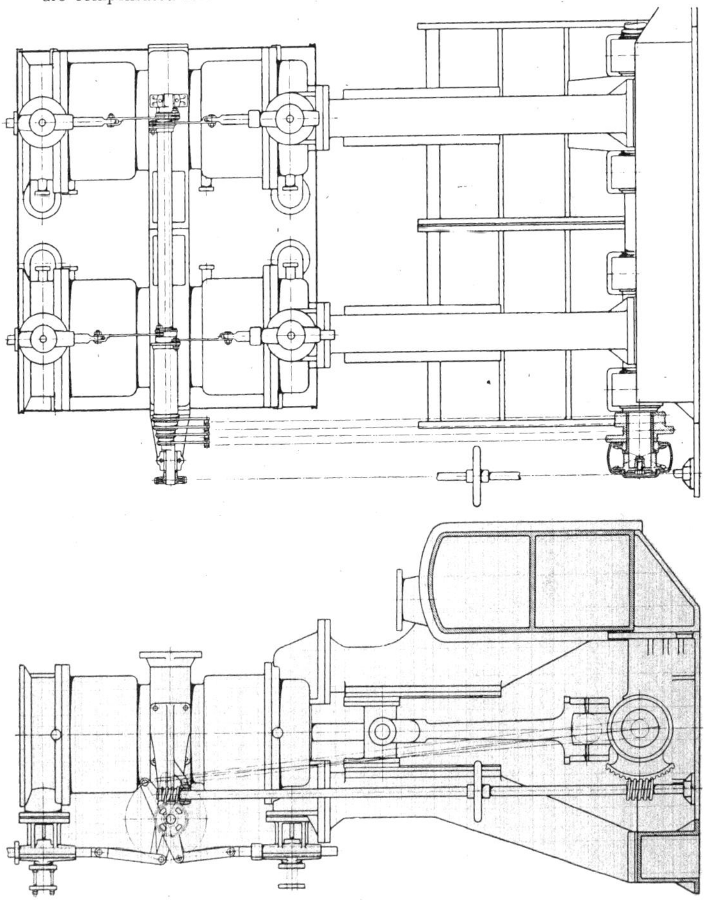

Fig. 241.

All the eccentrics are so disposed that each of the oscillating arms for the same cylinder moves in the opposite direction to the other.

An interesting modification of this gear is shown in figure 241. In this case, the eccentrics drive four valves through telescopic shafts. The telescopic shafts are arranged on a spindle which may be swung by means of a crank, the crank being supported in two bearings at the ends of the spindle. The crank, carrying the telescopic shafts, and the bevel wheel casing of the reversing gear are coupled by means of worm gearing so that the centre of the telescopic shafts is adjusted simultaneously with the main eccentrics. When the main eccentrics are moved from the minimum forward angle of advance to the minimum rearward angle of advance, the same adjustment takes place in the position of the common axis of the telescopic shafts. The throw of the main eccentrics and the arm of the crank on which the telescopic shafts are supported are so proportioned that the steam lap of the inlet valves is adjusted in conformity with the alteration in the angle of advance of the eccentric.

In this case, as well as in the valve gear described above, there is no necessity to provide any auxiliary valve, as with this form of valve gear, the cut-off may be made as much as 90%. The diagrammatic representation of the valve gear just described is shown in figure 242.

The principle of the valve gear shown in figures 240 to 242 is similar to that of the engines described with reference to figures 59 and 60. Also in marine engines it was important to obtain a proper lift of the valves during normal cut-offs without getting an unfavourable motion of the valves during later cut-offs of about 70 to 80% of the stroke.

The left hand valve diagram in figure 243, is for gears such as the Klug or Walchaert gear etc., whilst the right hand diagram is that for the above described gear. The right hand diagram is drawn so that the constant throw of the eccentric is equal to the maximum throw in the case of the left hand diagram.

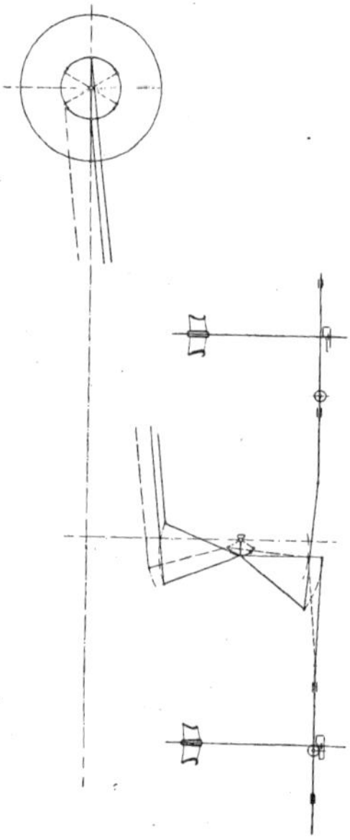

Fig. 242.

The valve opening A in the left hand diagram at latest cut-off is equal to the valve opening A in the right hand diagram at the latest cut-off.

During normal running however, when the cut-off is early, the valve opening in the left hand diagram is given by the amount B, whilst that in the right hand diagram is also similarly denoted by B. A glance at the figures will show that the valve opening during normal running in the case of the above described gear is much greater than that obtained by the Walchaert or other gears.

Figure 244, plate VI, shows a side elevation and plan of a una-flow marine engine for a paddle steamer plying on the river Volga in Russia. The cylinder diameter is 600 mm, stroke 800 mm. The power is 180 HP at 26 revs. per minute. Figures 245 and 246 show the cylinder casting of this engine.

Inclined rods of square section are arranged between the main shaft bearings and the cylinder covers and these rods form, over a part of their length, the cross-head guides. The main shaft bearings are cast in one with their supports. The valve gear is all assembled at one side of the engine and two eccentrics are employed. The valve gear is of the Klug type and operates a hollow and a solid shaft supported transversely in bearings on the top of both cylinders. The two depending arms of the eccentrics are connected at intermediate points by rods to a yoke piece

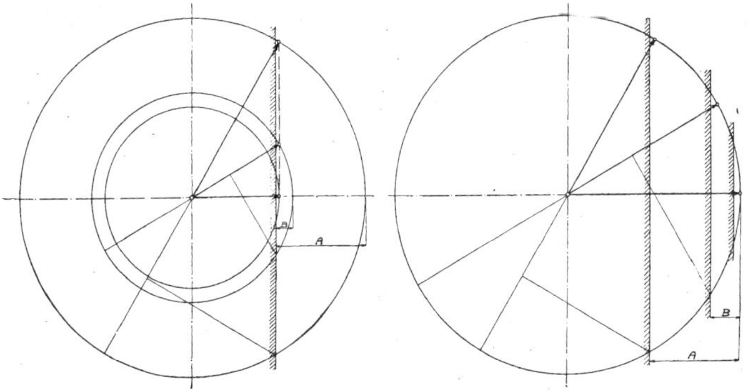

Fig. 243.

which may be adjusted by hand through screw gearing. The two main valve gears are designed for a maximum cut-off at 25%. The cut-off for normal operation on both sides of the cylinders is approximately equal. In the present case the auxiliary gear is coupled up to a different point of the short depending eccentric rods. In the construction illustrated, this auxiliary gear is connected to the eccentric rods at the same point as that to which the adjusting yoke is coupled. This auxiliary gear gives a cut-off up to 90% of the stroke, and thereby permitts of easy manoeuvring. An auxiliary rock shaft with a rock lever is supported in front of the cylinders, and the auxiliary slide valves are operated from this rock shaft. The auxiliary gear is cut out by suitable mechanism operated from a hollow shaft mounted on the auxiliary rock-shaft. This cut out gear is so coupled to the main operating gear that the entire control during manoeuvring is effected from the main gear.

Figure 247, plate 7, shows a modified form of this gear in which the telescopic shafts are supported on a swinging crank as described above with reference

Table VI.

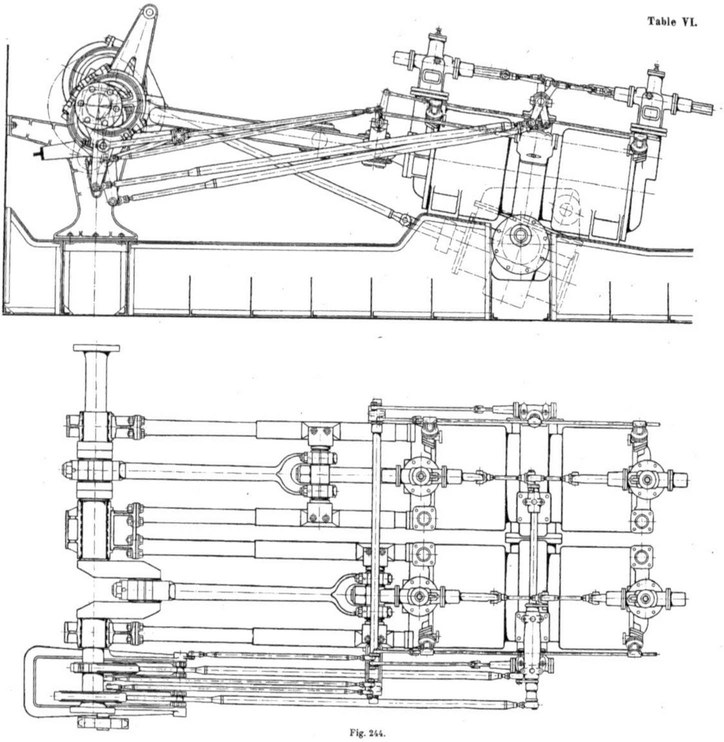

Fig. 244.

Printed by R. Oldenbourg, Munich and Berlin.

Table VII.

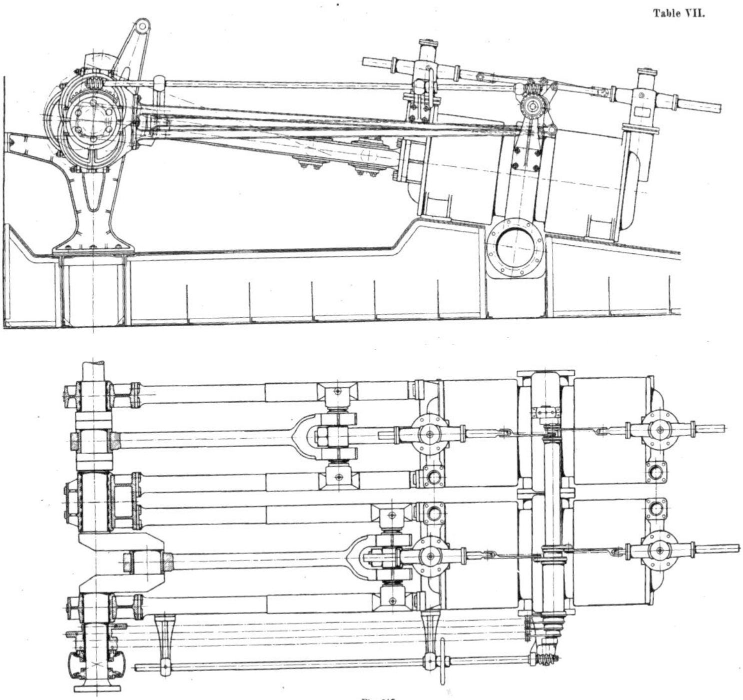

Fig. 247.

Printed by R. Oldenbourg, Munich and Berlin.

to figure 241. These telescopic shafts are arranged transversely between valve bonnets on the upper side of the cylinder.

In the case of larger powers when it is desired to use the Schlick balancing arrangement, four or six una-flow cylinders with a corresponding number of crank motions may be provided. A four cylinder engine of this kind is shown in figure 248. The hollow piston can be made very light, as was pointed out with reference to the locomotive engine. Light pistons should be employed in the outside units whilst additional weights can easily be applied on the inside pistons without in any way interfering with the remaining part of the construction.

Stresses on the working parts due to the inertia of the masses are much better distributed in the case of the una-flow engine than in the case of modern three and four stage expansion engines. In modern triple and quadruple expansion engines, the inertia and steam pressures have to be added together in the latter part of the

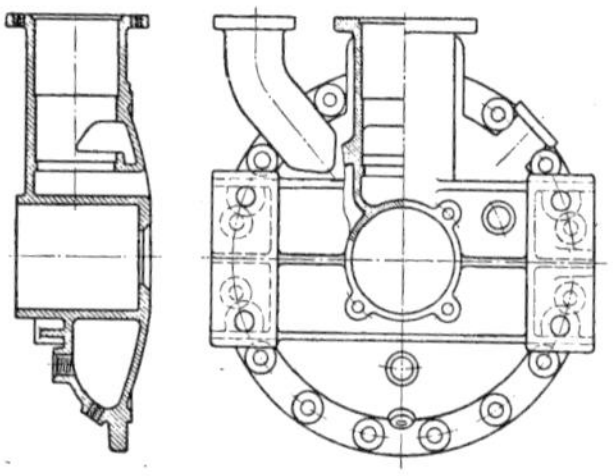

Fig. 245.

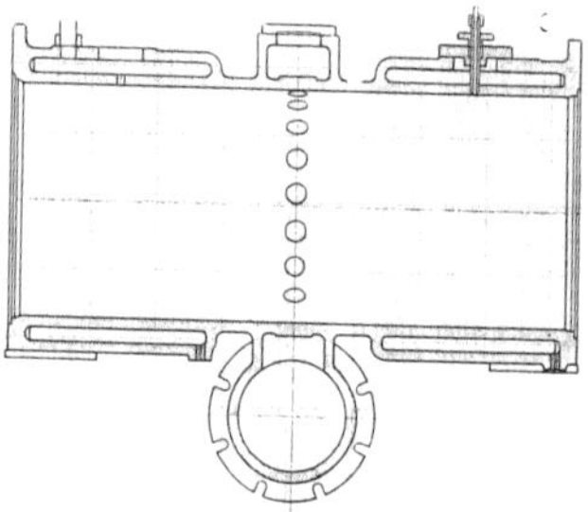

Fig. 246.

stroke. The maximum stresses, during running, actually obtained in large una-flow engines, are much smaller. The piston velocity may at the same time be 5½—7 metres per second. Investigation proves that at all greater loads, the stresses on the working parts are most evenly distributed in the case of the una-flow steam engine, whilst these stresses are better distributed in the multi-stage expansion engine of smaller units and at lesser speeds.

For lower and medium loads and lower speeds, the triple una-flow marine engine has many advantages such as smaller maximum stresses on the working parts, a more uniform turning moment and lower stresses on the crank shaft. With three, four and more cylinders, the una-flow marine engine offers a great reserve of power as each engine is a complete unit in itself and any single engine requiring repairs may be cut out whilst the remaining engines may be arranged to supply the additional power necessary to make up for the engine or engines which has or have been cut out. Taking an example, it will be appreciated that in the case of a four cylinder engine as illustrated, two engines can be cut out and the cut-off in the remaining engines could be increased from the usual 10% to 20% of the stroke thereby entirely making up for the two engines cut out.

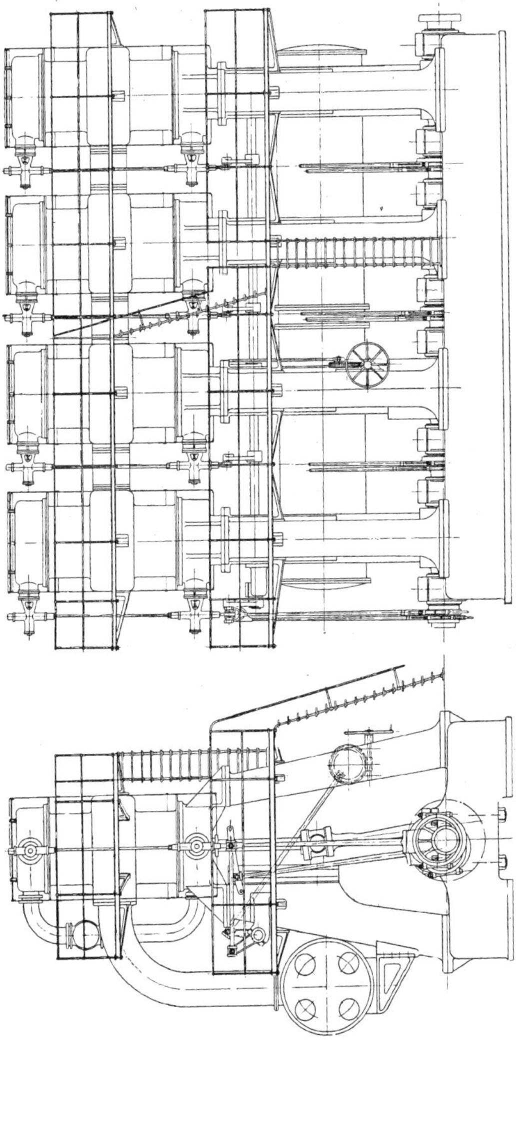

Fig. 248.

The reversing gear of the una-flow engine, is very much simpler than that required for multiple stage expansion engines. Only the inlet valves require to be manipulated during reversing. Then again in the una-flow engine, there are no intermediate receiver pressures which have to be taken into account when reversing, and moreover as the compression in the una-flow engine is always constant, the

Fig. 249.

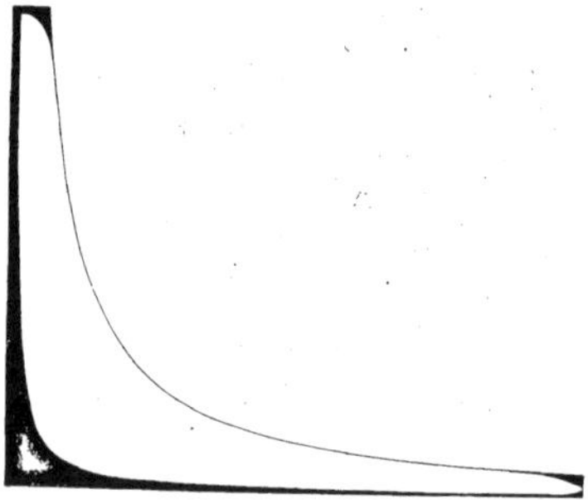

Fig. 250.

difficulties caused by excessive compression pressures in the first cylinders of multiple expansion engines, are avoided.

With a una-flow engine, the balanced lift valves, used for controlling the inlet, offer very little resistance, so that reversing is thereby greatly facilitated. In practice, it has been found, that in marine engines, as in locomotives, the valve gear is subjected to very little wear.

In a four stage expansion engine, the completeness of the diagram may be taken, as shown in figure 249, from 60% to 65%. The remainder is lost by throttling in the distribution valves, intermediate pipes, and by condensation losses. In the una-flow engine as shown in figure 250, the completeness of the diagram may be as high as 80% with a good vacuum, that is to say a difference of 15 to 20% in favour of the una-flow engine. The black part of the diagrams, represent the loss, whilst the white parts represent the actual work obtained from the steam in the engine. This also accounts for the material difference in the effective volume of the una-flow cylinders in comparison with the low pressure cylinder of a multiple expansion engine. This is also, to some extent, the reason for the fact that the steam consumption in a una-flow steam engine is not greater than that of a multiple quadruple expansion engine of the same power both for saturated and superheated steam.

By dividing the steam flow over several cylinders, the inlet valves are smaller as compared with the large dimensions of the slide valves usually employed in modern multiple expansion engines, where one great flow is performed through all cylinders in series.

Another important feature of these engines and one which is of considerable value is that all the spare parts may be applied to any one of the engines. This enables the number of spare parts carried to be greatly reduced, and is possibly of greatest value in the case of two cylinder engines. In the case of engines with more than two cylinders it would be possible to dispense with all spares for more than one section of the shaft in view of the great reserve of power possible in the engine itself.

Conclusion.

As will be gathered from the above Chapters, the una-flow engine is capable of general application. There is probably not a single branch of steam engine practice in which the una-flow steam engine could not be economically used. A brief consideration immediately points out the lines of development for any particular purpose. The difficulties usually relate to temporary or permanent reduction of the compression pressure. The main result of experience has been that multi-stage expansion and superheating can only be considered as unnecessary complications for most power purposes. It is, of course, also possible to work with multiple stage expansion and superheat on the una-flow principle. Experience has shown that highly superheated steam can be used with excellent efficiency and with greater reliability in the una-flow steam engine than in multiple expansion engines. The extraction of energy from the steam in stages and superheating are seemingly false lines of development consequent upon the use of the counter-flow system. Let the counter flow system be replaced by the una-flow and thereby obtain a cylinder with the utmost possible constancy in its thermal conditions; in other words substitute for the ordinary counter-flow cylinder a really adiabatic cylinder and then you will raise the reciprocating steam engine, from a thermal point of view, to the same high pinnacle, as that on which the steam turbine at present stands. The result of the transference from the counter-flow to the una-flow system is, that thermal simplicity replaces thermal complications and enables the same economic efficiency to be obtained without the necessity for multiple-stage expansion or superheating.

Spotted something that isn't quite right?

Let us know and we'll fix it for future reprints.
And as a thank you for your diligence, we'll
send you a free Sothis book of your choice.

publisher@sothis.press

This page intentionally left blank